PROSPECTORS, PROMOTERS, AND HARD ROCK MINERS

Tales of the Stewart, B.C. and Hyder, Alaska Camps

By

IAN McLEOD

With HOLLY McNEIL

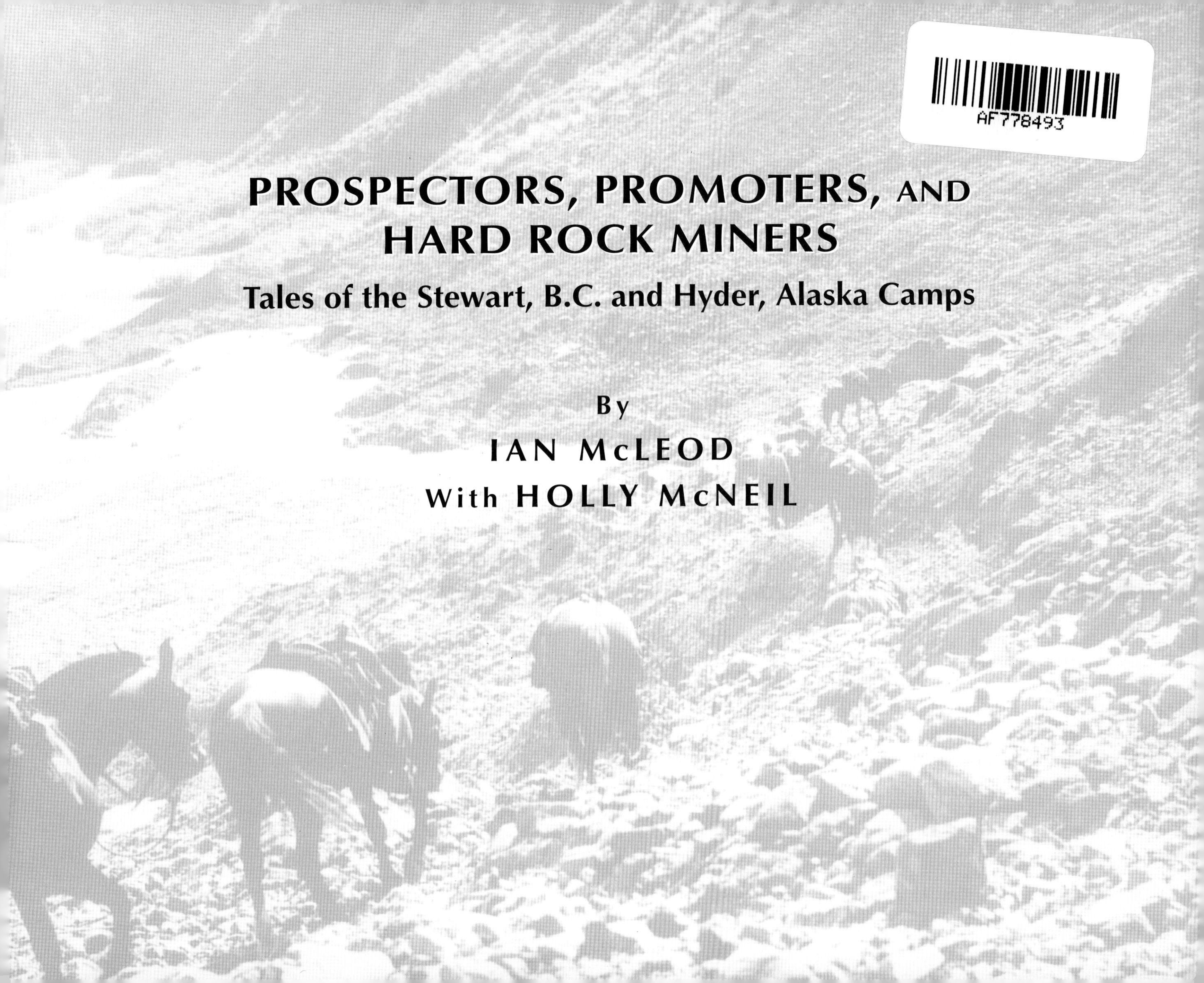

Dedicated to the pioneers
of the Portland Canal district.
They blazed and built the trails
for we who followed.

PUBLISHED BY
S.H. CO. LTD., 134 - 609 TRUSWELL ROAD, KELOWNA, B.C. V1W 3Z1

SALES AGENT
THE BOOK SHELF, 3250A BERRY RD., LAKE COUNTRY, BC V4V 1Y6
PHONE 250-766-2079
FAX 250-766-0643
EMAIL BOOKSHELF@PARAGONBC.COM
WEBSITE www.paragonbc.com/stewart

Acknowledgements

A special thank you to my family for their encouragement and inspiration on this project. To my wife Vivian, daughter Nancy, son Robert, brother Don and sister Janet Miller, I am deeply indebted for your help.

To my project partner and friend, Holly McNeil, a big thank you. Your help and advice has been first rate. Holly is a descendant of Portland Canal pioneers as is the author.

For photos and information, special acknowledgements go to:
Jack and Joyce (nee Stewart) Bouzek;
Paul and Miriam (nee Haahti) Goud;
Louise Skovcard (nee Behnsen), the great grand-daughter of
Bill Crawford;
David and Wanda Green;
Bill Walker.

Many thanks to Bill Ross (The Old, Bold Scot) for poems and advice. Also to Paul Wojack and Doug Flynn for advice and information.

Much appreciation to Bob Hutchings for jogging the old memory! Bob is the only one left who has a mountain named for him (Mount Hutchings) but at the age of 80, I doubt if he can still climb it!

To all above I will be forever grateful.

Ian McLeod

FIRST BRIDGE OVER BELL-IRVING RIVER, 1968.
STEWART-CASSIAR HIGHWAY #37

Contents

THE VILLAGE

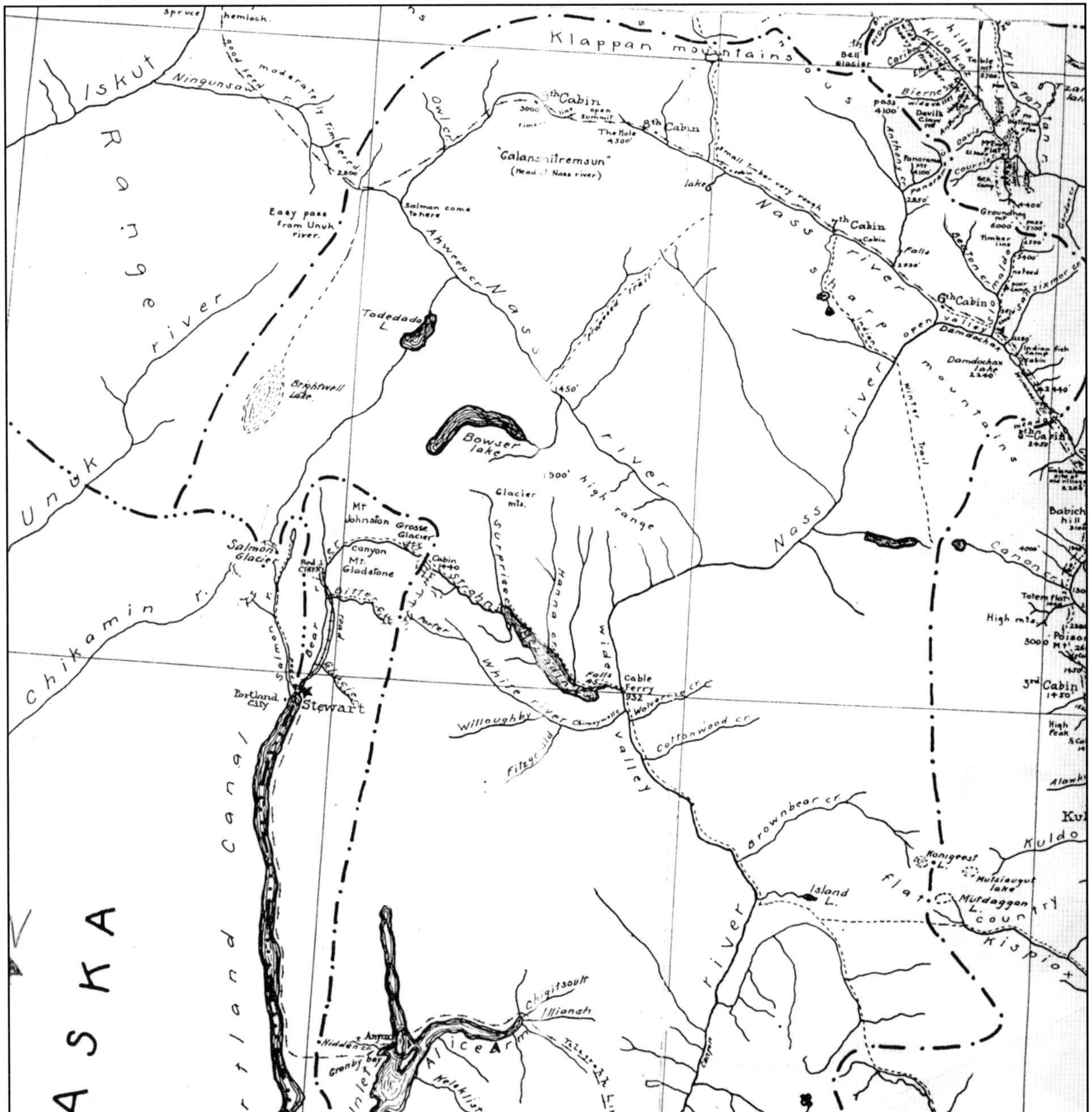

1910 MAP OF AREA. NOTE HYDER IS CALLED PORTLAND CITY. PRESENT BELL IRVING RIVER IS CALLED NASS UPPER BOWSER RIVER AND TIDE LAKE NOT SHOWN. FAR RIGHT IS YUKON TELEGRAPH LINE.

Early History

Chapter One

EARLY HISTORY OF THE PORTLAND CANAL

THE HEAD OF THE PORTLAND CANAL WITH HYDER TO THE LEFT AND STEWART ON RIGHT. THE SLASH UP THE RIDGE OF MOUNT DOLLY IS THE INTERNATIONAL BOUNDARY. THE SMALL SLASH TO ITS LEFT IS THE PREMIER TRAMLINE AND TO THE EXTREME RIGHT IS THE OLD POWER LINE FROM THE MARMOT RIVER TO SILVERADO MINE ON MOUNT RAINEY.

The history of the Portland Canal district has been dramatically shaped by its physical surroundings. Residents are influenced daily by tremendous amounts of snow and rain in a land of majestic mountains and sprawling glaciers, at the mouth of Canada's most northern ice-free port. And always, they are guided by the promise of economic prosperity from an abundance of nearby natural resources.

The Portland Canal district is subject to very heavy snowfall, with an average of seventeen feet in Stewart and Hyder, 55 feet at Premier Mine,

65 feet at Big Missouri Mine and an amazing 80 feet at the Granduc Mine mill site. In the winter of 1966/67, Granduc measured a near record-breaking 1,106 inches of snowfall.

One of the natural phenomena affecting the district is called isostatic rebound. Ten thousand years ago, the area was covered by ice several miles thick. The weight of the ice on the earth's crust pressed into the molten magmas. As it melted, the load lightened and the crust has been slowly rising, resulting in a high tide line on the delta that advanced 540 feet between 1904 and 1927 and is still advancing.

When I was a boy the high tides completely

STEWART WITH ROAD TRESTLE TO DOCKS NEAR PORTLAND CITY, ALASKA. TRESTLE ABANDONED WHEN STEWART-HYDER ROAD WAS BLASTED OUT OF THE SIDE HILL 1930.

CHARLIE WALKER WITH PACK TRAIN ON 5TH AVENUE. BUILDING ON LEFT WAS JOHN HAAHTI STORE, NEXT CRAWFORD OFFICE, THEN BANK OF MONTREAL. CRAWFORD LATER MOVED HIS OFFICE TO WHAT IS NOW MARMOT RIVER TRADING STORE AND THE OFFICE HERE BECAME THE NORTHERN ARGONAUT NEWSPAPER AND LATER MARIE BURNETTS CLOTHING STORE.

covered the main street (Fifth Avenue) and now the highest tides only reach hundreds of yards to the south. My geologist son, Robert, explains that marine gravels (named thus because they contain salt water shells) at Clements Lake in the Bear Valley and 11 Mile in Alaska are now 500 feet above sea level.

There are several large ice fields in the district that have been receding in recent years. The path of Highway 37A through the Bear Pass was covered by several hundred feet of ice as recently as the late 1960s, and remnants of the old road can be seen on the hills above the present highway that was constructed in 1973.

The area was first mapped in 1793. As Captain Vancouver was surveying the northern coast of British Columbia, he anchored his ship "Discovery" for a time in Observatory Inlet, some 70 miles south of Stewart and Hyder. From the reconnaissance point, he sent smaller boats forth on mapping expeditions up the various inlets. On that occasion, he named the 60-mile-long northward arm, the Portland Canal, after the Duke of Portland.

In 1896, Captain Gaillard of the U.S. Army Engineers, surveyed the canal and constructed four masonry building in the area, one of which still remains at the border in Hyder. (Capt. Gaillard went on to fame as an engineer during the building of the Panama Canal, where the Gaillard

◀ A GROUP OF LOCALS ATTENDING CHRISTMAS CONCERT AT THE OLD LEGION HALL ON SIXTH AVENUE.

D.J. RAINEY'S CABIN WAS BUILT IN 1899 AS PART OF A HOMESTEAD THAT MIGHT HAVE BEEN IN ALASKA OR IN BRITISH COLUMBIA. AS THE EXACT LOCATION OF THE BORDER HAD NOT YET BEEN DETERMINED, "DAD" RAINEY HEDGED HIS BETS AND REGISTERED HIS HOMESTEAD IN BOTH COUNTRIES. THE LAND IS NOW RAINEY CREEK PARK IN STEWART, AT THE END OF 7TH AVENUE. BOTH RAINEY CREEK, WHICH RUNS ALONGSIDE, THE PARK, AND MOUNT RAINEY IN THE BACKGROUND WERE ALSO NAMED FOR "DAD." HE WAS A PROSPECTOR WHO STAKED THE DISTRICT'S VERY FIRST CLAIM, ABOUT SIX MILES UP BITTER CREEK, AND NAMED IT THE GRIZZLY. DURING A PROSPECTING TRIP ON MOUNT DOLLY IN HIS LATER YEARS, HE FELL FROM A CLIFF AND DIED.

Cut is named for him). As he was the first to survey the area, Gaillard established a boundary line up Bear Valley and, at the 56th parallel of north latitude, he constructed a cairn which still exists today.

The next record of visitors to the district is 1898 by a con man named Burgess. He circulated the story that he had discovered a placer field at the head of the Portland Canal, one as rich as the Klondike. Burgess chartered the ship "Discovery" in Seattle and sold berths to 68 men who believed him. (It is only coincidence that the ships captained by both Vancouver and Burgess had the same name). So the second "Discovery" offloaded the men and their supplies on the Bear River tide flats and steamed away. After a day or two of organizing their gear, Burgess declared that he would lead the men upstream to the new placer ground. Instead, in the dark of the night, he slipped away on the lone remaining small boat, never to be heard from again. This was the first mining swindle of our history but it certainly wasn't the last.

These abandoned men started prospecting anyway, but the gravel yielded only traces of gold. However, several of them had lode prospecting experience and discovered mineral in the solid rock. One such person was D.J. Rainey, who located the first claim to be recorded in the area, the Grizzly Claim on Bitter

THE PORTLAND CANAL SHORT LINE RAILWAY ON THE STEWART WATERFRONT IN 1910. REMNANTS OF THE PILINGS CAN STILL BE SEEN.

Creek. Rainey also pre-empted a homestead on the west wide of Bear River and recorded it under Alaskan law as it was presumed that the boundary would run up the center of Bear Valley. when, in 1902 the official boundary was established between the Bear and Salmon valleys, the Canadian government honored Rainey's pre-emption, and his name was later affixed to two Stewart landmarks, Rainey Creek and Mount Rainey.

Brothers John and Robert Stewart, founders of the town, also arrived in 1902. They staked and surveyed the land east of Rainey's and began selling lots. By 1910, several lode properties had been identified and staked. The Portland Canal and the Red Cliff outfits decided they had enough ore to commence production but they required transportation. Enter Sir Donald Mann and his partner Mackenzie, who had just finished building the Grand Trunk Railway to Prince Rupert, and were looking for new worlds to conquer. They formed the Portland Canal Short Line Railway, sold stock in the venture, and commenced construction. By the end of 1910, they had build sixteen miles of track to the Red Cliff

CRAWFORD TRANSFER FLEET IN FRONT OF NEW OFFICE. LARGE BUILDING WAS NEVER FINISHED AND WAS DEMOLISHED FOR LUMBER FOR MORRIS SUMMIT MINE. BUILDING BESIDE IT WAS MOVED TO 7TH AT VICTORIA AND IS NOW OWNED BY DAVE AND WANDA GREEN. THE LAST BUILDING WAS OWNED BY JOHN BOUZEK AND WAS THE POST OFFICE IN THE FORTIES AND FIFTIES.

◀ THE PORTLAND CANAL SHORT LINE RAILWAY STATION IN 1910. ITS LOCATION IS NOW CALLED RAILWAY AVENUE.

mine and were able to begin hauling out ore. But that didn't last long: when the limited tonnage made continued operation infeasible, the promoters closed down that line and cooked up a plan to raise money for a rail bed extension to the Groundhog coalfield and eventually to the Peace River, where they would haul grain to an ocean port. But the investing public wouldn't bite, and the project was in limbo until 1914 when the rolling stock of engines and flatcars were shipped out to Vancouver to be used in the war effort. The rails were kept in place until 1941 when they, too, were used elsewhere in the World War II effort.

THE STEWART TOWN SITE, 1910, WITH THE RAILWAY IN THE FOREGROUND AND MOUNT DOLLY BEHIND. ➤

The Portland district is the quintessential "boom and bust" community. Established in 1905, the population reached several thousand by 1910. However, over the next few years, the lack of a major mineral discovery dissuaded many, and they left in droves. A 1932 edition of the *Stewart News* reported that by 1917, the population of Stewart was 52 and in Hyder, only 37.

After the discovery of the rich ore by Premier Mines in 1919, the two towns boomed again and until 1930 there was a great deal of building and prosperity with a district population of about 1,000. The beginning of the Depressions slowed investment in mining but, by 1935, with several hundred men employed at Premier and a similar number at Big Missouri mines, the district survived the 'hungry thirties' better than many other communities in North America.

War in 1939 caused an exodus of young men who joined the armed forces and by 1946, the population had dipped again to only 300 people in Stewart and about 50 in Hyder. In 1951, the Granduc copper discovery combined with the start of local development in the timber industry influenced another gradual climb. In 1956, the government of Prime Minister John Diefenbaker commenced construction of the Stewart-Cassiar highway, which was designed to haul asbestos to a deep-water port in Stewart. That road was finally completed in 1971 and a year later, the Nass River Bridge was opened by the Hon. Ray Williston, providing road access for the Portland district to the rest of the world. It was a long campaign to convince the government of the need for access, and Stewart residents Mary Schindel and Jim McKelvie were tireless in their lobbying efforts.

By 1970 there were more than 2,000 people in the district. But Granduc's closure in 1983 began another downward spiral so that today, in 2003, there are fewer than 600 people in the area. Logging has replaced mining as the primary industry, with summer tourism and transportation also contributing significantly to the local economy. The former Granduc deep-sea bulk loading facility now handles shipments of gold and silver ore from Eskay Creek and copper concentrate from Huckleberry Mines (located near Houston, B.C.).

The great accessible glaciers and rugged mountains attract many tourists in season. The grizzly bear viewing area near Hyder records more than 20,000 visitors annually.

Stewart and its citizens have never given up, and now with road access to a developed deep-water port, I predict that they never will. Former resident and author Ozzie Hutchings summed up the local spirit with total accuracy when he titled his 1976 book, *Stewart, the B.C.-Alaska border town that wouldn't die.*

CRAWFORD TRANSFER FLEET ON FIFTH AVENUE, 1935. ON THE FAR LEFT IS THE POST OFFICE AND NEXT TO IT, NEWELL ROOMS, BAYVIEW HOTEL, TOOTH SMOKE SHOP, MELLOR STORE AND THE MARMOT HOTEL.

WITH THE PORTLAND CANAL ADJACENT, FISHING WAS A FAVOURIE SPORT FOR RESIDENTS. THIS HAPPY CREW ALL GOT A HALIBUT EACH. L TO R: JOHN LUPICK, ROY STANICH, RUDY KASUM, NICK LUPICK, CHARLIE JACOBSON, AND GEORGE LUPICK. PREMIER ORE BUNKER BEHIND. ➤

HIGH TIDE ON FIFTH AVENUE. 1925.

FIFTH AVENUE, 1910. NONE OF THESE BUILDINGS LEFT STANDING.

Mines and Miners

Chapter Two

PACK HORSES AND SNOWSHOES

CRAWFORD HORSES PULLING GO-DEVILS ON THE WAY TO GEORGIA RIVER GOLD MINE IN 1928

During the 1920s and '30s, the heyday of mining exploration in the Portland Canal district, pack horses were the transportation of choice for most miners. There were hundreds of mining prospects, and dozens of pack horse trails leading to them.

More than a hundred horses were located here and in the off-season they ranged freely throughout Hyder and Stewart, especially on the tide flats where grass grew belly-high on the horses. For a few months every winter, the stock had to be fed in the barns as they were not able

RENNIE AND YOUNG PACK TRAIN
ON SALMON GLACIER
1930

to scrounge grass under the five or so feet of snow that typically fell in this northern coastal climate. They were joined by a dozen or so dairy cows, owned by John Campbell and later Bob Hutchings. With all these grass cutters, the villages sometimes looked like a park — except for the cow-pies, which we kids had to dodge repeatedly while horsing around!

The two biggest horse operators were Young & Rennie, and Crawford Transfer. Jack Rennie of the former and Bill Crawford of the latter were both top-notch men — hard, smart workers who

AMERICAN MINING AND MILLING PACK TRAIN LEAVING HYDER, ALASKA FOR THE MINES AT FISH CREEK 1910

PACK HORSES CROSSING AMERICAN CREEK 1925

looked for similar characteristics in the packers they hired. In the '30s, when brother Don and I were growing up, horse packing was considered to be a very glamorous profession, and we loved to hang around the barns. In return for shoveling them out, we were permitted to learn to ride, saddle up, horseshoe and to throw the diamond hitch, which was used to hold the load on top of the horse. Eventually, we grew experienced enough to work with the packers.

In 1919, Bill Crawford imported a big mule from Alice Arm. He called the mule 'the Blue Ox' undoubtedly after Paul Bunyan's famous big oxen, Babe. The mule could carry a much heavier load than a horse and was used to pull big pieces of machinery. Mules, however, were much more difficult to handle than horses so the Blue Ox was the only one used in the Portland Canal area as far as I know. Bill would say that it was essential for a good mule skinner to have a slightly higher IQ than the mule!

From the packers, Don and I learned a good trick about feeding horses on the trail. If the camp was above timberline, we had to have the horses carry their hay but down in the valleys, there was lots of grass and oats beside the trail. (The oats were seeded by the undigested oats in the horse manure.) There were also numerous blue grouse along the trail that fed on the oats. So, when we stopped at night to set up camp, we would unpack and un-harness the horses so they could pasture during the night. The problem, of course, was that the horses loved their home barn and pasture, and if they had a chance they would head there at a gallop. One solution was to find a place at the home end of the trail where you could block their way with a tarp held by ropes. But the best way was to drape a bearskin, instead of a tarp, on the ropes because horses have an inborn fear of bears and will never voluntarily come within sight of a bearskin. The next challenge was to transport the bearskin – you had to wrap it in a tarp before you could load it on the horse's back. (Even in the saddle one had to be alert in bear country because at the first scent, a horse could spook and perhaps throw the rider.)

In the early days of Portland Canal packing, the wooden saw-buck pack saddle was widely used. When the U.S. Army Decker saddle became available, they quickly switched. The Decker was quite superior; it was all leather and metal, and much easier on both the horses and the men.

The horses could pack their loads in a number of ways. On their backs, they could carry up to 300 pounds, but that wasn't nearly enough.

A "go-devil" sleigh was devised to be pulled, usually over dry, fairly level trails. The go-devil was about two feet wide and three feet long. The two runners were six-inch saplings with a bend in the front. Dragging behind this little sleigh was a deck of planks onto which the freight was fastened. The go-devils were used in both winter and summer, but the runners didn't last very long on bare ground. Fortunately, there were lots of saplings growing on the side hills for replacements.

One of the most important elements of packing a load to be dragged was to ensure that the

➤ RENNIES PACK TRAIN ON SALMON GLACIER, 1930.

HORSES IN TANDEM HAULING TANK ON A GO-DEVIL ➤

GEORGE COPPER 1930

sleigh would not run into the horse's hind legs on a downhill pull. The animal simply wouldn't endure it and would invariably run away. To solve the problem, the packer would attach short loops of chain over the front of the runners to create drag. These chains were called "rough locks."

The popular go-devil did not work well to transport tonnages of high-grade ore down the

A HORSE EATING FROM A BALE OF HAY ON HIS LOAD. TO PREVENT SUCH FREELOADING, MOST PACKERS WRAPPED THE BALE IN A TARP. THESE ARE WOODEN SNOWSHOES.

steep slopes of the mountains. In its place, the packers used a system called "rawhiding." The green hide of a cow was filled with ore sacks at the mine, sometimes as much as one thousand pounds. The hide was laced up with rope and hooked to a horse in harness. Because it was downhill, the horse was able to pull the heavy load; because of the green hide, it did not slide into the horse's heels. A rawhide pack was only good for one or two trips before it was worn through but replacements were cheaply available from the slaughterhouse.

Carl Jones, one of the last of the rawhiders in the Portland Canal, liked to tell a story of his favorite horse, Dolly, who was once pulling a load from the United Empire mine. The promoter, Billy Dann, was walking wearily behind and decided to ride for awhile on top of the load. Dolly had a habit, when reaching particularly steep sections, of flopping her rear end onto the

HORSE'S SNOWSHOE
THIS WAS KNOWN AS THE CRAWFORD TYPE SHOE AND WAS DEVELOPED BY BLACKSMITH LADE WHO WAS A BROTHER-IN-LAW OF BILL CRAWFORD.

CRAWFORD TEAM ON SNOWSHOES HAULING BOAT FROM BIG MISSOURI MINE TO LONG LAKE TO BE USED DURING POWER DAM CONSTRUCTION.

load and steering with her front feet. Unfortunately for Billy, he was sitting right behind Dolly and she flopped directly on top of his legs! Although nothing was broken, it was several weeks before he could walk properly again. (Another Billy Dann story: he once played host to a visiting party from the Vancouver Board of Trade who had come to town on the CNR steamer. Bill barbecued a bear and broke out several kegs of beer. The party became known as the Three B Party, for Bear, Beer and Bullshit!)

Another very helpful packing tool, particularly in the Portland Canal area where average winter snowfall was usually between five and 12 feet, depending on the elevation, was the horse snowshoe. The snowshoe was certainly not developed here: Robert Scott on his South Pole expedition in 1906 used Siberian ponies with snowshoes. I saw a pair from the 1800s in a museum in the Scottish highlands: they were made of oak and iron and looked quite clumsy but were apparently used over both snow and muskeg.

The metal snowshoes used in the Portland Canal district were built completely in the forge as there was no electric or acetylene welding in those days. Frank Hough built wooden shoes with two pieces of three-quarter inch plank, nailing and clinching them together.

Although horses have been used underground throughout the ages, I only know of one time they were used in the Portland Canal district. They pulled ore cars at the Big Missouri Mine in the 1930s and were later retired to pasture in Stewart. Crawford owned the two that I remember, Snap and Missouri, and he loved his animals — he was not one to reward a lifetime of hard work with a bullet behind the ear as others have done. An old miner told me of his shift boss who must have also been an animal lover. The boss told the crew: "There is too much gas underground for the horses. This morning, boys, you'll have to hand tram." (To tram was to push the ore cars to the surface. The blasting created noxious gas that normally was blown out of the workings by fans.) He was clearly more concerned about

SNOW AND MUSKEG SHOE CIRCA 1700 AT MUSEUM IN SCOTLAND.

➤ ORE SACKS PILED ON FIFTH AVENUE. BUILDING WITH PITCHED ROOF WAS POST OFFICE. STILL STANDING AND USED AS BAKERY.

OWNEY MCFADDEN WITH A TRAIN CROSSING AMERICAN CREEK IN 1946.

➤ HOMEMADE SNOW PLOW MADE FROM AN OLD ROAD GRADER, 1925.

the horses than the condition of the men's lungs!

In 1930, Crawford negotiated a winter freighting contract at Barkerville, B.C., which had started to boom due to ore discoveries by Fred Wells. Bill took 30 head of horses to Barkerville where he built a barn and house, but he returned to Stewart when the job was finished. His big job during the '30s was freighting for the Big Missouri Mine that was getting ready to go into production. At the end of the decade, the daily rate was $5.00 for a horse and $5.00 for a packer. No discrimination!

The last pack train operator was Owney McFadden who kept going until the mid-1950s when the arrival of the helicopter finally put the horses out of business. Owney was an Irishman from Donegal, and he lived up to the blarney reputation so well that his nickname was Bullshit McFadden. In 1938, he reported to the Stewart detachment of the provincial police that a meteorite had fallen at the Bear Pass and that the wind from the impact was so great, it knocked over his tent four miles away. Larry Requa was the investigating police officer, and he related that there was lots of evidence of the impact. In fact, he said that the little round lake in the gravel moraine beside Highway 37 at the east end of Bear Pass was where the meteorite must have hit. But... Larry also told a story about finding a cave with three alien skeletons while searching for a lost trapper on the Tide Lake flats. I knew that story wasn't true as my friend Happy the Trapper was with Larry on that trip, and Happy told me it was a lot of bull!

➤ BILL STEWART WITH A HORSE ON WOODEN SNOWSHOES.

▼ A SIX-HORSE TEAM ON FIFTH AVENUE, 1925.

MCMILLIANS STAGE LINE STEWART TO BITTER CREEK ➤ 1910. EMPRESS HOTEL BEHIND.

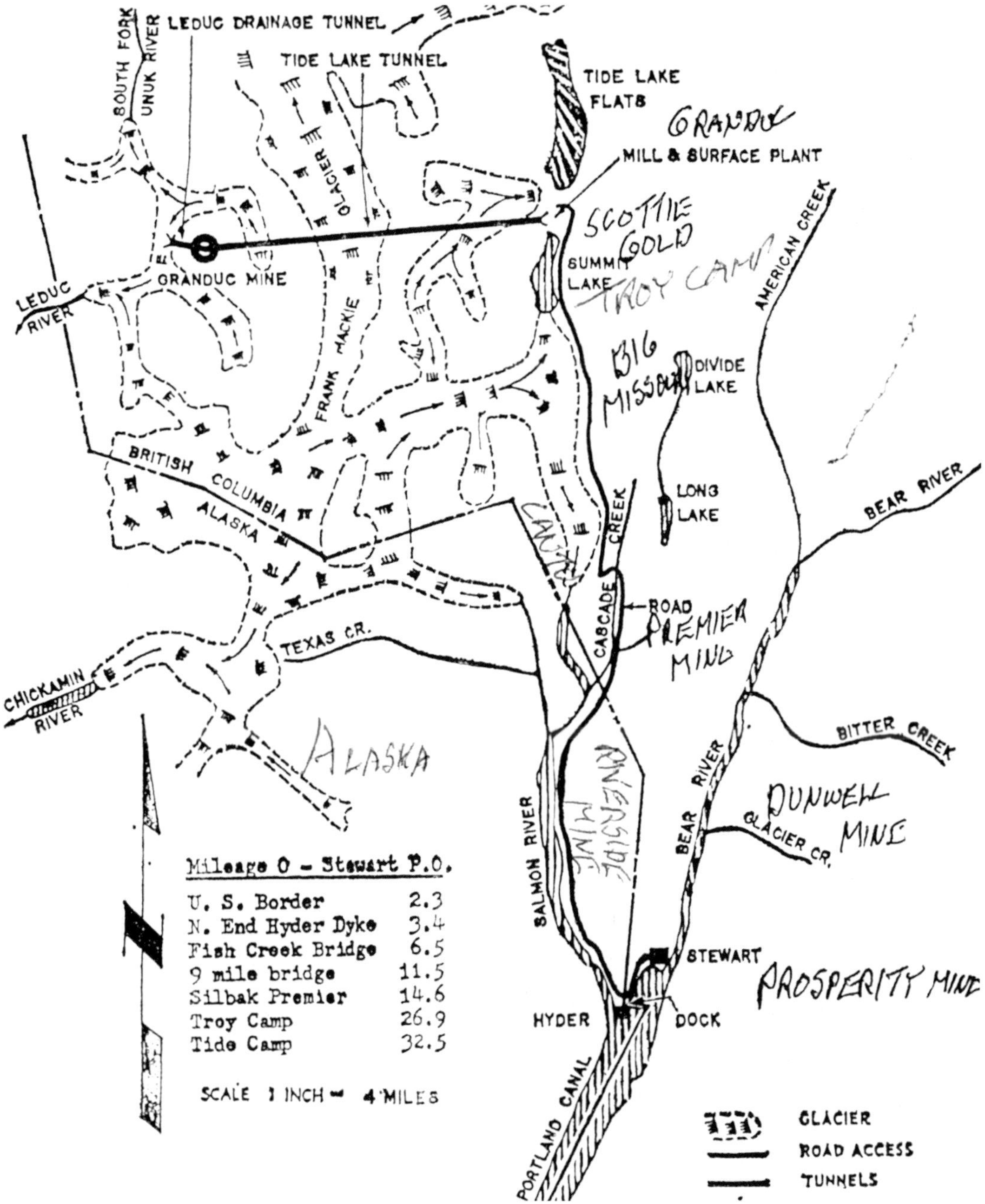

THIS MAP SHOWS THE LOCATION OF GRANDUC AND ITS LONG TUNNEL.

Chapter Three

GRANDUC MINE

The development of the Granduc copper mine in the 1960s changed the course of history in the Portland Canal district. Although never a profitable mine, it nonetheless represented extraordinary accomplishments, led to one of B.C.'s most tragic disasters, and resulted in a remarkable legacy for the town of Stewart.

The discovery of the ore body was made by two people, one of whom was my friend Tommy McQuillan. Tommy told me: "In 1951, I was prospecting for Karl Springer Helicopter Exploration and we were checking properties in the Unuk valley. My partner was Einer Kvale and the field geologist was Don Cannon. I suggested that we go to the Leduc valley to check a copper showing. The helicopter only took two passengers, so Einer and I went first. We landed at a large gossan beside the Leduc Glacier. Don joined us on the next trip and after looking at the ore and the size of the showing he told Einer and I to start staking!"

Springer optioned the property to Granby Mining, which formed a subsidiary named "Granduc" — a name derived from merging Granby and Leduc. (Incidentally Karl Springer, a successful miner, later financed Russ Baker in the formation of Pacific Western Airlines.)

In 1953 diamond drilling began at the Granduc property and by the early 1960s, they had blocked out 40 million tons of 1.7 per cent copper. Workers accessed the site by aircraft landing on the glacier and also by Cat train via the Premier mine road and over the Salmon Glacier. Some thousands of tons of freight were moved to the mine during those years at the cost of two deaths: one man fell into a crevasse while checking the depth of ice to determine if the tunnel could be driven in solid rock under the glacier. The other victim was a Cat skinner whose D-7 caterpillar slipped into a large crevasse. His body was recovered but the D-7 is still somewhere in the ice.

Granby's first partner was Newmont and later Asarco joined to form Granduc Operating Co., which put up the $150 million needed to begin production.

That was a very tough job. The challenge was to transport ore under two glaciers for eleven miles to the processing plant. A tunnel was the only solution, and the resulting Granduc tunnel was a world-class effort and accomplishment.

GRANDUC MILL, 1980.

The tunnel was begun simultaneously at both ends, with a plan to meet in the middle, but in February 1965 an avalanche of incredible proportions wiped out the western portal camp, killing 26 men. (See the chapter entitled Granduc Slide.) The result of that disaster was to close the western portal, and drive the tunnel strictly from the east.

The mill site itself was 32 miles away from the nearest port at Stewart, which presented another significant challenge. Road construction was commenced in 1965 and the project included a one-mile vehicle tunnel that was driven under a dangerous avalanche path. When the road was

GRANDUC MILL SHOWING DAMAGE CAUSED BY SNOW, 1979.

completed, focus was on development of a 5,000 ton per day mill, a power plant and a camp. Single men lived in bunkhouses at the mine site and although first class recreation facilities were also built, the annual employee turnover in the first few years was more than 300 per cent of the 800-man work force.

A town site for married workers was built at Stewart and these employees travelled daily to the mine by bus. Part of the road was avalanche prone but a strict safety program was established and when the mine closed, the road still had a perfect safety record.

Production commenced in 1970 and ceased in 1978. During these years, the project was barely profitable and managers continually bemoaned the price of copper. Many times I heard the refrain: "If we only had 75 cent per pound copper, we would be successful."

In 1979, Imperial Oil decided to get into coppering and formed Canada Wide Mines, which bought out Granduc. The big mill roof had collapsed in the short interim from the formidable snow load but it was repaired and production

DECEMBER 8, 1968 PARTY TO CELEBRATE BREAKTHROUGH OF 11 MILE GRANDUC TUNNEL. "ON LINE AND ON GRADE."
L TO R BACK ROW: JOHN MOROZ, TUNNEL BOSS; UNKNOWN; MAYOR IAN MCLEOD; NICK GRITZUIK, VICE PRESIDENT, GRANDUC; MERV UPHAM, VICE PRESIDENT, NEWMONT; UNKNOWN; DON HOWARD, MANAGER, GRANDUC; HARRY BAPTY, BC MINES INSPECTOR

recommenced. Managers told me that company economists had predicted that copper prices would soon rise to $1.00 per pound but that price had not materialized by 1982. The mine was shut down that year and the plant totally dismantled. Today, road access to the mine site remains but there is little sign of the once bustling community.

Two miles north of the mine camp was the Tide Lake flats. On the empty lake bottom there, the company built a 1,000-foot airstrip that was used to ferry freight before the road to Stewart was complete. Adjacent to the airstrip was a camp of 20 mobile homes where some married couples lived as well as the single female employees. The idea was to protect the ladies from the several hundred men housed in the bunkhouses at the main camp! These men, out of envy no doubt, nicknamed the small camp "Happy Valley!"

Beyond the airstrip was an area where the cookhouse garbage was dumped. Feeding two

BOSS GRIZZLY "EMPEROR" AND FRIEND AT TIDE CAMP GARBAGE DUMP 1967.

or three hundred men created a lot of garbage and before long, the local grizzly bears took over the dump. The bears were focused strictly on the food and didn't bother the garbage man, who soon had named his favorites. He called the boss grizzly, "Emperor." At one time, there were 22 regular garbage customers, and a Vancouver nature photographer had a field day photographing them. He always made sure his images didn't include the garbage! The photographer also produced a film entitled "The Valley of the Grizzlies" with no mention of garbage. When the refuse was finally buried, the bears had to go back to their old hunting habits.

The Granduc ore handling facility is at the Stewart port. It is presently owned by Stewart Bulk Terminals, with Stewart resident Al Soucie as the managing partner, and trans ships ore from Eskay Creek mine and Huckleberry mine near Houston, B.C.

Granduc is gone but it has left quite a legacy for the town of Stewart. The huge surge in population during the 1970s necessitated development of a sewer and water system, housing, road paving and recreational facilities, as well as a deep-sea bulk loading facility that is now fully operational. Also, the road to the mine site is one of the most accessible glacier viewing areas in the Pacific Northwest.

ARROW TRANSFER CONCENTRATE TRUCK ON THE GRANDUC ROAD

The following poem was written by a group of participants in the tunnel project and was given to me at the grand opening celebration of the tunnel breakthrough on Dec. 8, 1968.

Midst the glaciers and the mountains,
In the Province of B.C.,
Newmont and Asarco,
Have made mining history.
Now they had undertaken
What had been tried before
To drive a tunnel, ten miles long,
And mine the Granduc copper ore.
The cost to drive this tunnel
And develop the Granduc mine
Was eighty million dollars,
And would take three years of time.
The success of this development,
To the financiers, created no fears,
They formed Granduc Operating Company,
And leased the property for seventy years.

The camp was then established,
At a place they named Tide Lake,
Thirty miles from the nearest town.
But a good place to save a stake.
So word got around about the tunnel,
That Granduc was going to drive
The rumors claimed the money would be good.
So the crews began to arrive
Italians, Canadians and Germans,
English, French and Dutch,
Welshmen and Yugoslavs, Hungarians and such,
Irishmen and Norsemen and Scots there were a few,

GRANDUC LOADING FACILITY IN STEWART HARBOUR. IT IS NOW OWNED BY AL SOUCIE AND JACK ELSWORTH AND ORE FROM ESKAY CREEK AND HOMESTAKE MINE IS SHIPPED HROUGH THE DOCK.

They came from all parts of the world
To form the Granduc crew.

Early in this project,
Unlikely did it seem
That from this motley crew of ours,
Would evolve a 'Highball' team.
But patience, trial and error
Helped us find our way
And soon were advancing
Seventy feet a day.
But then our rapid progress,
Came to an instant halt
At distance eighteen thousand feet,
We hit an argillaceous fault.
Experts recommended shot Crete, perfo
bolts and net,
These were not successful, so we introduced,
Operations Steel Set.
A laborious, tedious, monotonous job,
But one that had to be done.
And five months later we were happy to find,
Our problems overcome.
Then spring brought a manpower shortage
Seemed that no one wanted to stay,
The only one happy at this stage of the game
Was that airline P.W.A.

Drill, blast then muck her out.
Rock bolts if the ground was poor,
Hanging pipe and laying track,
There was a lot of work for sure.
In the execution of this work,
Competition between shifts arose,
The crews were led by Old Timers three,
Solonkia and Doskotch and 'Duster' Moros.
Miners were occasionally AWOL,
Not of sickness, but usually tight.
The reasons when given were always the same,
'A little too much snakebite.'
Disciplinary action was then enforced,
In the form of a Warning Slip.
Copy for Office, copy for Miner and copy for
the 'Whip'.
Slip one was usually accepted in fun,
And good for a laugh and a smile,
But Slip number two would earn for you,
One day off and your name on the file.
The next offence was three days off,
And number three on the slip,
If you stepped out of line one more time,
You'd be making a farewell trip.

In spite of all our setbacks,
There passed another year,
And our performance in the tunnel assured,
That completion was drawing near.
And then one day it happened,
Something to make our success complete,
We set a One Day World Record Advance of
115 feet.
This was the crowning achievement,
The fulfillment of a dream,
We had earned the coveted record,
By being a 'Highball' team.

I wish to broach an issue
Which I think is worthy of note

To inform you how our leisure is spent,
In our location so remote.
We have Snooker and there's Billiards,
Chess, Draughts or a show,
Curling, Darts and Rummy.
So who cares about the snow?
But what we really need to overcome our plight
Is for Personnel to send us some Ladies of
the Night.
Blondes, Brunettes and Redheads
Olive, Black or Dusky Tan.
These women skilled in their profession,
Would east the tension of any man.
I think that this suggestion
Is in keeping with the times.
That women should be permitted
To work around the mines.

In the final days of our Project,
Engineering confirmed Line and Grade, true.
On December 8th, 1968,
We officially broke her through.
We have gathered here to celebrate,
A job that was well done,
A performance where records were made.
That will stand for years to come.
I wish to leave you with this thought,
In my closing line,
This tunnel will stand as our monument
Until the end of Time.

- *author unknown*

TIDE CAMP PORTAL WITH BERENDON GLACIER BEHIND. THE TUNNEL WENT UNDERNEATH THE TWO GLACIERS ON RIGHT.

Chapter Four

THE GRANDUC SLIDE

It was exactly 10:16 a.m. on the morning of Thursday, Feb. 18, 1965. There was more than five feet of fresh snow on the ground at Portal Camp beside the Leduc Glacier, and it was still snowing heavily. Many of the workers were heading to the cookhouse for a coffee break when, without more than a whisper of a warning, a wall of snow crashed into them.

The avalanche, later described as one of the worst in Canadian history, buried 50 of the 140 men in camp, destroyed eight of the ten buildings, and temporarily devastated the mining operation. Twenty-six men lost their lives.

Members of a drilling crew that were working in the tunnel just downhill from the camp were the first on the scene. The impact of the slide had affected their underground electricity feed and the compressed air source to the tunnel, so the men ran to the surface immediately. When they looked out, the powerhouse, machine shop, the helicopter and its hangar, and the garage next to the portal had vanished. They scanned for evidence of a second slide before they ventured forth with their shovels and headed for camp. By then, some of the victims had dug themselves out and other unhurt survivors were digging wherever they heard something.

By a quirk of fate, Dr. C.V. Veasy of Stewart and Fred "Doc" Moore, the much-liked senior first aid attendant, were in camp that day and unhurt by the avalanche. They established a makeshift hospital in the undamaged office building. At the same time, Innis Kelly, the young Pacific Western Airline weather and radio man, was cobbling caterpillar batteries together to enable the sending of a mayday distress signal.

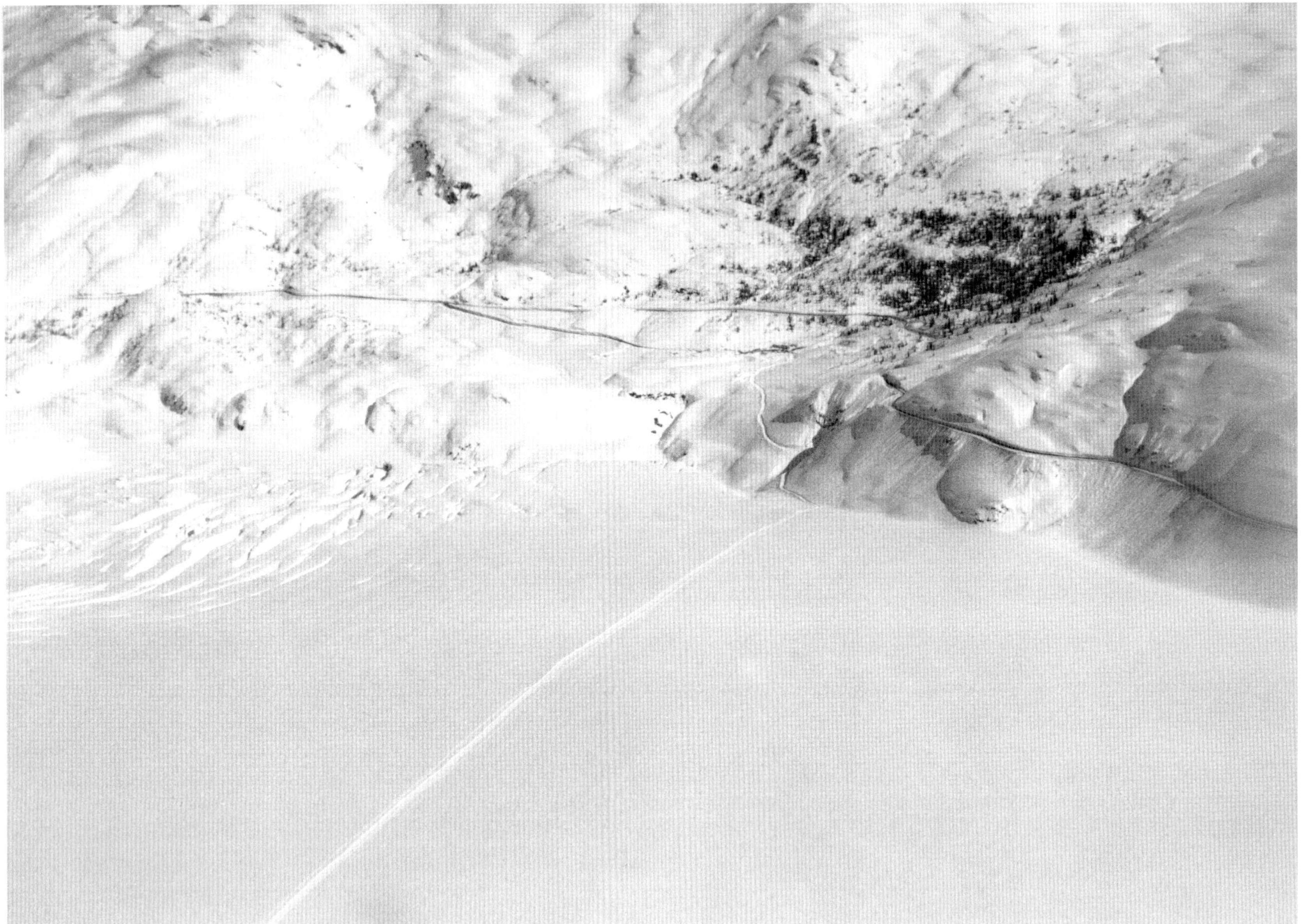

THE GRANDUC ROAD AT TROY CAMP, 1965. THE CAT ROAD ON THE SALMON GLACIER IN THE FOREGROUND IS THE ROUTE THE RESCUE CATS TOOK AFTER THE AVALANCHE.

SALMON GLACIER. CAT TRAINS IN WINTER CAME OVER THIS GLACIER FROM CANTU ON THE LEFT AND TROY CAMP ON THE RIGHT. THEY HEADED OVER THE SUMMIT ON THE HORIZON AND DOWN TO LEDUC GLACIER AND PORTAL CAMP.

The signal was received in both Ketchikan and Stewart, and immediately the whole north country on both sides of the border sprang into action. Among the first to mobilize were pilot Kenny Eichner of Ketchikan and Dr. Jim Wilson, who headed out together in a Temsco helicopter. The weather was terrible as they flew toward the mouth of the Chickamin River, which they intended to follow upstream to the Leduc River and then on to Portal Camp. By the time they arrived at the toe of the Leduc Glacier visibility was so bad they had to land the helicopter. But they were able to contact Portal Camp by radio, and the miners said they would send a Tucker Sno Cat to pick them up. Unfortunately, the Sno Cat had the same visibility problem and tipped over sideways on a steep side hill. The driver and his helper were unhurt but they had to return to camp, leaving Kenny and Dr. Wilson stuck in the helicopter until morning.

Meanwhile, the rescue operation was proceeding in numerous different manners. Helicopters from the whole north country were being readied, the Ketchikan rescue squad was preparing to strike out at dawn, and the Vancouver rescue squad was on its way north. Governor Bill Eagan of Alaska diverted the state ferry "Taku" to Stewart and the Canadian vessels, "Nanaimo" and "Laurier" set out to join the U.S. Coast Guard vessel "Romaine" at the mouth of the Chickamin. It was announced that 50 soldiers from the Royal Canadian Engineers were on their way.

The big freight Cats were at Troy Camp waiting for the weather to lift when they heard the Mayday and they immediately headed across the Salmon Glacier, attempting to reach the 5000-foot summit before going down to Leduc Glacier. The snow was so blinding that one of the operators, a very tough man by the name of Freddy Hassilberg, tried to go ahead on snowshoes and give the Cats a trail but they finally had to give up.

By dark on that Tuesday night, there were still 28 men missing. The survivors were cold and hungry. The injured — six seriously, eleven less so — were housed in a makeshift hospital, patched up as best as Dr. Veasy and "Doc" Moore could manage. A big bonfire was fuelled easily enough with wood salvaged from the ruined buildings, and the miners scavenged cans of food and some frozen steaks from under the wreckage of the cookhouse. It was difficult to get close enough to the towering bonfire to cook the food, but they solved the problem by using their long-handled snow shovels as frying pans. Steaks were soon sizzling, and canned peas and corn were eaten with their fingers. Pete Overwater was an old miner notorious for complaining about camp grub, no matter how fine it was. He was finishing his steak when one of the mine cooks commented: "Pete, this must be fine food. It's the first time you never complained."

Pete replied: "Shut up and put another steak on the shovel for me."

It was not until dawn on Friday that the weather cleared enough for the helicopters to land at camp. Because the pass to Stewart was still socked in, the injured were flown to the mouth of the Chickamin and transferred to the waiting boats or, in the case of the seriously injured, directly to Ketchikan Hospital. At 10 a.m., the rescuers successfully dug out Clarence Moore who had been trapped in what was left of the tool shed. That left 27 missing.

The Cat train under boss Johnny O'Shea got moving early Friday, too, but conditions were still difficult. The big Cats didn't arrive at Portal Camp until Saturday at 4 a.m. and they immediately went to work. The rescue technique was to cut six inches of snow at a time with the guidance of two point men riding on each corner of the bulldozer blade. Whenever the point men thought they saw a body, they would signal the operator to stop, and they would dig with shovels. Al Soucie and Bob McKay, both of Stewart, were operators there and told me about the situation. Soucie said that they quickly learned to tell when they were close to a body as the snow around it would be crumbly and dark. By Saturday, hope of finding any more survivors was slim, and Dr. Veasy returned to Stewart.

But Sunday morning, a miracle took place! Operator Roger McPherson had just made a sweep with his D-7 when one of the point men saw an arm move under the snow. They immediately started digging and realized they had a conscious man. Soucie and McKay jumped down to help get the man onto a stretcher. The victim's first words were "give me a shot of whiskey!" McKay remembers that when they got him to the makeshift hospital the man, whose name was Eino Myllyla, complained about his hip. McKay lifted the fellow's hip and found a big ball of ice frozen to his clothes. After 79 hours (more than three days!) buried under the slide the man was flown immediately to Ketchikan where Dr. Wilson took charge. Myllyla had frozen hands and feet, and it was determined that a hyperbaric chamber was needed. The closest one was in Buffalo, New York so it was also flown to Ketchikan.

PORTAL CAMP MARCH 1965. CAT TRACKS SHOW WHERE THE SEARCH FOR BODIES CONTINUED.

Myllyla spent 78 hours in the chamber that was filled with pure oxygen and pressurized to three atmospheres. It worked, and several weeks later he was released from hospital. But he took to the bottle with a vengeance and two years later was killed in a car accident along with a partying friend.

After Myllyla was rescued, the operation went into a recovery mode that lasted for months. When the slide had stopped its violent downward motion, it left only eight feet of snow at the campsite but 30 feet at the portal where bodies and debris were piled up. The last two bodies were not recovered until June and as I knew both the men I was called on to identify them. Not a pleasant task.

My friend Jimmie Warner was driving a Bombardier Sno-Cat when the slide hit, and he was buried for an hour in the cab. He blew the horn every few minutes until someone dug him out. Jimmie gave up camp life after that experience and worked for me for 20 years. Glen Sullivan, a married Stewart man with six children, was in the cookhouse when the slide hit, and he survived with a broken back and leg. Two of my former mining buddies were also lucky enough to survive the onslaught: Malcolm "Porkchop" Campbell was under the snow for 30 minutes before he dug himself out; and Cy "One Eye" McLellan was safely working in the tunnel. However, Cap Palmer with whom I used to work was killed. Cap happened to be on the dump at the portal when the slide came down. I believe he was the only underground miner lost. All the other deceased were carpenters and laborers who were working on camp construction.

As a result of the slide, the Granduc Company hired a world-famous avalanche expert who developed a safety plan for both Portal Camp and the road from the mill to Stewart, which travelled through a large number of avalanche tracks. With mortars and bombs and good men, the mine never again experienced an accident due to avalanche.

The original plan to drive half of the eleven-mile tunnel from Portal Camp was abandoned after the slide, and the tunnel was driven completely from the east end (see chapter on Granduc).

But Portal Camp was repaired and the miners working from there prepared the ore body for production until the tunnel broke through and underground access from Tide Camp was possible. Then Portal Camp was dismantled.

If anything, the Granduc disaster demonstrated what many of us already knew for fact — that Northerners are of the same family and regardless of which side of the border we reside on, we work together for the common goal. Of the eleven rescue choppers, six were from Canada and five from Alaska.

Because the word 'avalanche' was not in the miner's every day vocabulary, I have used their term – slide. Even today when they talk about it, they call it the Granduc slide or the Big Slide.

Chapter Five

PREMIER MINE

11 MILE ON THE PREMIER ROAD IN 1920. AN AVALANCHE CAME DOWN ON THE ROAD SO THE MANAGER CLOSED THE UNDERGROUND AND DIVERTED THE MINERS TO TUNNEL THROUGH THE SLIDE. THEY SHOVELED THE SNOW ONTO THE SLEIGHS. DURING THIS PERIOD, RICH RAW ORE WAS HAULED TO TIDE WATER AT STEWART USING FOUR HORSE SLEIGHS. CRUDE ORE IN 1919 AND 1920 AVERAGED 4.24 OUNCES OF GOLD PER TON AND 141 OUNCES OF SILVER.)

The Premier mine was responsible for the prosperity of Stewart and Hyder during three decades beginning in about 1918. Its production record of two million ounces of gold and 45 million ounces of silver makes it the third most productive precious metal mine in British Columbia's history. The Bralorne-Pioneer camp produced 4.7 million ounces of gold followed by the Rossland camp with 3.2 million. The biggest silver producer was the Sullivan at Kimberly with 223 million ounces with the Premier second with 41 million ounces. The Eskay Creek mine (see Chapter 14) sixty miles northwest of Premier has already beat the Premier ounces and I predict before long it will be known as the richest gold and silver producer in B.C. ever.

First staked in 1910 by Scotty Dilworth (Mt. Dilworth) and the Bunting brothers (whose name is on Mt. Bunting), it was surrounded by other claims owned by numerous partners and investors. Dr. W.T. Kergin and Dr. Richards of Port Simpson each had a claim given to them by Scotty Dilworth as payment for medical services when Scotty broke his leg. Of all the small claimholders, the only one I knew personally was Jake

1920 SIX-HORSE TEAM HAULING FREIGHT TO PREMIER. SACKED ORE WOULD BE CARRIED ON THE RETURN TRIP. WHEN PREMIER CONVERTED TO HAUL ORE SLEIGHS WITH CATERPILLARS THEY FOUND THAT THE CATS HAD A HARD TIME GOING AROUND THE SWITCHBACKS AS COMPARED TO THE SIX-HORSE TEAMS. SO HORSES WERE USED ON THE SWITCHBACK SECTION AND ONCE ON THE STRAIGHTAWAY THE CATS TOOK OVER THE SLEIGHS. IT MUST BE REMEMBERED THAT BULLDOZER BLADES WERE NOT YET INVENTED.

Hawkinson who was presented with one claim for providing grub to the prospectors.

In my youth, Jake operated the Portland Canal Electric Laundry. He later declared he never got a nickel for his claim. Farther up the "hill," McKenzie Logan and John Morrison staked a group of claims that would become the main ore-bearing claims of Premier. 'O.B.' Bush formed a company and all the claims were amalgamated into the Cascade Falls Mining Company. O.B., as he was called, was from Edmonton and had only one arm due to a childhood accident. He had managed, however, to become a world-class speed skater in his youth.

O.B. hired Pat Daly as superintendent for various prospecting and development activities but the early results were not rich enough. In 1917, Pat happened to be in Spokane and told R. K. Neil, an engineer, about the property. Neil traveled to Stewart to size up the project, then contacted Messrs. Trites, Wood and Wilson of Fernie. They, along with Neil, optioned the property from O.B. & Associates for $115,000. Under Neil's

FIRST PREMIER BUNKHOUSE. BURNED 1928.

➤ PREMIER CAMP 1935. MANAGER'S MANSION IS IN FOREGROUND ON RIGHT.

▼ PREMIER 4 LEVEL MILL, 1935.

direction, a tunnel was built and within a month they had hit high grade. In 1919, American Smelting and Refining bought 52 per cent for a million dollars and put the property into production.

The Gugginheim family controlled American Smelting and Refining. They were known far and wide as the "Guggies" and were notorious as "high-grade and run" operators. And run is exactly what they did when the ore grade began deteriorating in 1936, despite the $23 million in dividends produced up to that point. But the mine stayed alive under English control for many years as Silbak Premier.

By 1921, camps and facilities had been built and the tramline was operational. In the 1930s, when I first saw Premier, it was a self-contained town with a doctor, hospital, movie theatre, school and basketball facility, store, tennis courts and a seasonal swimming pool. It was, however, an isolated town during the six months or so of winter each year, until large snow removal equipment was utilized in 1936.

The mill ran 400 tons a day, and it was supported by a machine shop, a welding shop, blacksmith, and just about everything else needed to be self-sufficient. Although the wages were not very good, people were generally satisfied because they understood how the rest of Canada was suffering through the Great Depression.

Although Premier was mainly a gold and silver operation, the miners had also found lead and zinc on the lower levels. That was what they mainly produced during World War II.

In the '30s, Premier had about 200 employees with 150 living in the bunkhouse and about 50 families in various types of housing scattered on the hillside. There was a dining room and kitchen for the single workers and a separate dining room for the supervisors that was nick-named the "Blue Room" because apparently the supervisors always looked down in the mouth!

One of the many characters I knew at the Premier was Jack Hamer, known far and wide as Jackhammer. Jack was an old broken down miner, deaf as a door nail and like so many, very fond of the bottle. When a miner was beyond his hard working years the Premier company would

➤

SPORTS DAY AT PREMIER IN 1935 WAS HELD THE FIRST WEEKEND IN AUGUST. AS WELL AS CHILDREN'S RACES AND OTHER SPORTING EVENTS, IT FEATURED MINING COMPETITIONS LIKE THIS DRILLING ONE. THE DRILL MACHINE WAS A CRANK FED WATER LINER ON A TRIPOD.

➤

DIAMOND DRILLER JACK CAMPBELL WITH HIS DRILL IN 1938. JACK, WHO WAS MY COUSIN, STARTED HIS CAREER AS A HELPER AT PREMIER IN 1936 AND ENDED IT AS PRESIDENT OF BOYLES BROTHERS DRILLING, ONE OF THE LARGEST DRILLING COMPANIES IN THE WORLD AT THAT TIME. IN 1940, HE WAS SENT TO ENGLAND WITH HIS CREW TO DRILL HOLES UNDER THE BEACHES WHERE THE GERMANS WERE EXPECTED TO INVADE. THE HOLES WERE THEN FILLED WITH EXPLOSIVES AND WIRED TO EXPLODE WHEN THE ENEMY LANDING BEGAN. FORTUNATELY, HITLER'S INVASION DID NOT HAPPEN.

transfer him to an easier job. In Jack's case he became the King Nipper. The King Nipper was the man who transported supplies to the working miners. Explosives, sharp steel, etc was pushed on a flat car over the rails to the drifts, raises and stopes. Jack was very good at his job but at times he would go on a bender and miss a few shifts. Al Kirby, the underground superintendent had warned Jack a number of times that he would have to let him go if the benders didn't slack off. But Jack did it again. Al called Jack into the underground office and told him he was fired. Jack cupped his hand to his ear and said, "Yes, Mr. Kirby. I'll do that right away" and headed off down the drift. Al said later that although he fired a lot of men trying to fire a deaf one was just about impossible. So Jack stayed on until the mine closed.

Dale Pitt was the manager from 1921 until 1934 when he transferred to Australia to run the Big Bell mine. Pitt was a real autocrat as were most other managers of his era. He lived in a nice mansion on the hill with Japanese gardens and a Japanese gardener. Tom Lee, an old-time miner who lived his retirement years in Stewart, told a story about one Christmas when two drunken immigrants asked where they could find a bootlegger. Tom pointed at Pitt's mansion and the drunks headed that way. Several days later, Tom was called into the office and summarily fired. "Pitt just didn't have a sense of humor," he said.

In 1942, Pitt was in the States working for the American government on strategic war minerals. He was in charge of mining mica in New England and North Dakota. But he ran into a U.S. Senate Committee investigating the high cost of the

PREMIER 6 LEVEL MILL 1964

PREMIER MILL, 1995

mica and was hauled onto the carpet for a tongue-lashing. The head of the committee was an opinionated Senator named Harry Truman. Dale was apparently no match for Harry!

In 1952, Premier closed partly because of low metal prices but by 1956 things had improved and a crew was sent in to restore equipment to working order. A welder was repairing things in the mill when his torch ignited the oil-soaked timber and the mill burned to the ground. Many of us thought that would be the end of Premier but not so. Good mines die hard!

In 1958, my friend and schoolmate Julian Berkosia was prospecting in the former upper

glory hole of the Premier. Where a rock wall had earlier caved in, Julian discovered a new very rich vein. I always said Julian had a nose for ore and he proved it that time. With partners Tommy McQuillen and Bob Hutchings, they obtained a one-year lease from the company and commenced shipping raw ore. In that year, the partnership cleared a quarter of a million dollars, which was a lot of money at the time. The Premier company then commenced work on the new discovery and by 1964 had built a 75-ton mill at level 6. My brother Don was manager — a definite promotion from his earlier mucking days at the mine! In 1965, Don left for better things and the company of Bralorn-Pioneer took over management, bringing in Adam O'Krainec as manager. When Adam retired from the mine, he became my right-hand man at the King Edward hotel.

▲ PREMIER 4 LEVEL BUNKHOUSE. OFFICE ON RIGHT. 1963.

THE PREMIER CANADA CUSTOMS OFFICE WAS ONE HALF MILE ➤ FROM THE BORDER AND WAS CLOSED SHORTLY AFTER 1952. LAST CUSTOMS OFFICER STATIONED THERE WAS JACK BOUZEK.

A GOOD LOAD OF SNOW ON PREMIER 4 LEVEL DUMP. ADAM O'KRAINEC AND MIKE TINSLEY. 1968.

"THE BIG MUSHROOM" OIL STORAGE TANK 4 LEVEL., 1968.

ABANDONED 4 LEVEL BUNKHOUSE, 1968.

THIS 1923 TELEGRAM TO HARRY GUESS, THE VICE PRESIDENT OF AS&R COMPANY (THE GUGGIES) WHICH CONTROLLED PREMIER WAS FROM DALE PITT, THE MINE MANAGER. THE NORTHERN LIGHT GROUP WAS ADJOINING PREMIER AND WAS CONTROLLED AT THE TIME BY "TRUE BLUE MACKENZIE." PREMIER EVENTUALLY BOUGHT NORTHERN LIGHT AND DISCOVERED RICH LEAD AND ZINC ON THE PROPERTY. THE LEAD AND ZINC WAS MINED DURING WORLD WAR II AND THAT PRODUCTION ALLOWED THE MINE TO STAY OPEN AS GOLD AND SILVER MINES WERE CLOSED BY THE GOVERNMENT AS BEING NOT ESSENTIAL TO THE WAR EFFORT.

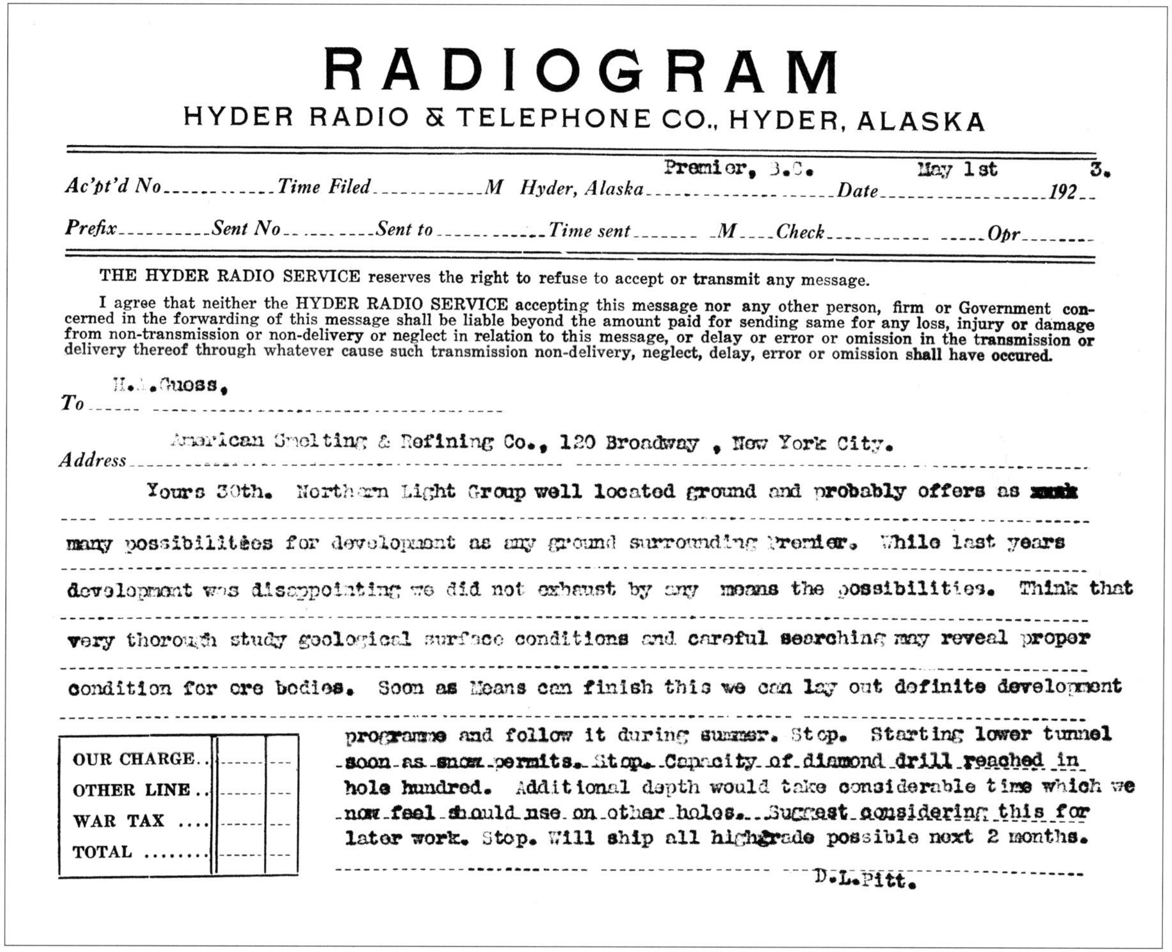

RADIOGRAM

HYDER RADIO & TELEPHONE CO., HYDER, ALASKA

Ac'pt'd No ____ *Time Filed* ____ *M* *Hyder, Alaska* Premier, B.C. *Date* May 1st *192*3.

Prefix ____ *Sent No* ____ *Sent to* ____ *Time sent* ____ *M* ____ *Check* ____ *Opr* ____

THE HYDER RADIO SERVICE reserves the right to refuse to accept or transmit any message.

I agree that neither the HYDER RADIO SERVICE accepting this message nor any other person, firm or Government concerned in the forwarding of this message shall be liable beyond the amount paid for sending same for any loss, injury or damage from non-transmission or non-delivery or neglect in relation to this message, or delay or error or omission in the transmission or delivery thereof through whatever cause such transmission non-delivery, neglect, delay, error or omission shall have occured.

To H.A.Guess,

Address American Smelting & Refining Co., 120 Broadway, New York City.

Yours 30th. Northern Light Group well located ground and probably offers as many possibilitèes for development as any ground surrounding Premier. While last years development was disappointing we did not exhaust by any means the possibilities. Think that very thorough study geological surface conditions and careful searching may reveal proper condition for ore bodies. Soon as Means can finish this we can lay out definite development programme and follow it during summer. Stop. Starting lower tunnel soon as snow permits. Stop. Capacity of diamond drill reached in hole hundred. Additional depth would take considerable time which we now feel should use on other holes. Suggest considering this for later work. Stop. Will ship all highgrade possible next 2 months.

D.L.Pitt.

OUR CHARGE		
OTHER LINE		
WAR TAX		
TOTAL		

In 1981, Westmin Resources took over the Premier operation and after years of diamond drilling, they decided to put the mine back into production by open-pit mining the whole hillside. In 1989, a 2000-ton mill was built and for seven years the ore from both Premier and the Big Missouri properties were milled there. From three million tons of ore, they produced 260,000 ounces of gold and five million ounces of silver.

Since the mid-1990s the mill has been in mothballs.

TRAM BOSS
JIMMY BURNETT,
1930.

A MAN RIDING THE ORE BUCKET ON THE UNITED EMPIRE TRAM THAT WENT FROM SEA LEVEL TO A 3500-FOOT ELEVATION WITHOUT A TOWER.

Chapter Six

AERIAL TRAMLINES

Due to the very high snowfall in the Portland Canal district, horses could not efficiently move large amounts of ore and freight so aerial tramlines were utilized at several mines. It was a system of cars on cables, held high above the ground so that snowfall did not hamper the operations. They were very similar to modern ski hill gondolas.

The tramline at Premier Mines operated over a distance of eleven miles. There were four cables in view, the top two about six feet apart and fixed at both ends. The weight of the moving cars rested on these cables. The two lower cables were attached to the cars, with the operating principle that the loaded cars going downhill would pull the empty cars — and some freight — - uphill by a continuous cable and a system of pulleys to make the corners. Gravity provided the power. This was similar to the household clothesline except the tramline was horizontal instead of vertical.

UNITED EMPIRE TRAM TERMINAL IN 1938.

THE STEEL TOWER ON THE PROSPERITY TRAM.

These old-timers were pretty smart. They used an electric motor to launch from a standstill but once things were underway, a switch would turn the motor into a generator and the electricity generated could be used for the mine. The Premier tram used a total of 49 miles of cable, with 250 baskets hauling 800 lbs each.

The United Empire mine, one mile above Stewart, ran their tram cables from the east side of Bear River up to the Mount Dolly timberline on the west side with a single span over a mile in length. The mine did not last long but the cables stayed up for many years until we had to blast and break them because they had become a hazard to aircraft landing on the Stewart airstrip.

The Prosperity – Porter Idaho property was located on the east side of Mount Rainey at a 5,000 feet elevation. In 1925, the Premier Company took it over and constructed a five-mile long tramline built with steel towers over which rich silver ore was shipped for two or three years. The total production contained more than two million

PROSPERITY MINE BUNKHOUSE, 1930.

FAR RIGHT:
THE PROSPERITY - PORTER IDAHO TRAM TERMINAL, 1930. THE TERMINAL WAS AT A 5,000 FEET ELEVATION AND IT WAS LOCATED FIVE MILES FROM SALT WATER.

ounces of silver but by 1930 it was shut down permanently.

Both the Dunwell mine and the Portland Canal mine had two-mile tramlines but neither were profitable and were not in use more than one year.

The Premier tram ran from 1922 until 1944 when snow removal equipment was able to keep the road open and trucks were used to haul the concentrate.

In 1946, the Premier company bought the Indian mine across the valley from Premier, and built a tramline to haul the ore to the Premier mill, about two miles in length. Unlike all the other trams in the area that did not carry people, the Indian tram buckets were designed to carry one man each. The system allowed the men to live at Premier and ride daily to Indian. These 'man carriers' were like stretchers with wheels above, and the occupant had to lie down in a box-like contraption. The men called them "flying coffins."

One day, the tram operator did a faulty passenger count and thinking everyone was back in Premier, he shut down the tram and went home for the night. One poor man was left in the last "flying coffin" 200 feet above ground, with no way out. Despite the fact they kept a couple of blankets in the box, the poor devil still spent a very uncomfortable night. In the morning, the starting shift was surprised to see a half-frozen person staggering out of the forgotten coffin. The negligent operator was tipped off and he left camp before the victim warmed up!

MIKE AND "RED" MULLEN "LINEMEN" ON TENSION 8 PREMIER TRAM.

NOTE THE WAY SNOW HAS PILED ON A TREE STUMP.
PREMIER TRAM 1935.

Chapter Seven

BIG MISSOURI MINE

THE BIG MISSOURI DAM OVERFLOWS IN 1960.

The Big Missouri Mine is located four miles north of Premier Mine, very close to the Alaska Panhandle but still in Canada. It was first staked in 1904, but development work did not begin until 1927 when the Consolidated Mining and Smelting Company (now Cominco) optioned the property.

It showed promise so the mine was put into production in 1938. An underground mill of 750 tons per day was constructed, and the mine was worked until 1941 when it succumbed to the wartime pressure that closed many gold mines.

BIG MISSOURI CAMP, 1938.

BIG MISSOURI POWER DAM UNDER CONSTRUCTION IN 1937.

CALLED THE SHAKE BARN BECAUSE IT WAS MADE FROM HAND SPLIT SPRUCE SHAKES. HALFWAY HOUSE TO BIG MISSOURI, 1929.

What was amazing then about Big Missouri, and still is today, is that this was an underground operation with a grade of 0.08 ounces per ton, which is less than one-tenth of an ounce of gold per ton. This is abnormally low, but was buoyed by a hydropower system installed by the company. From a dam at Long Lake, some 500 feet above the powerhouse, the water powered two 1700-horsepower Pelton wheels that were attached to electric generators.

Big Missouri produced 58,000 ounces of gold and 52,000 ounces of silver during its production period.

The area around the mine is quite picturesque during brief summer periods, and it has become quite popular with local hikers. Heather and other beautiful wildflowers cover the hillsides, and the lovely views include a number of small, jewel-like lakes.

POWER PRODUCTION IN 1927,

BIG MISSOURI POWER HOUSE.

BILL CRAWFORD WITH HIS UNDERGROUND PONIES "SNAP" AND "MISSOURI."

LONG LAKE DAM LOOKING WEST.

Chapter Eight

HYDER, ALASKA

HYDER 1928. BUILT ON PILINGS WITH PLANK ROADS.

Hyder came into being along with its neighbour, Stewart, in the early 1900s. Its original name of Portland City was rejected by the U.S. Postal Service when the first post office opened in 1915 because there were apparently too many other "Portlands" in the States. The Lindeberg brothers, who owned the town site, suggested the name after Frederick B. Hyder, a Canadian mining engineer who had examined and reported favorably on Lindeberg mining properties. Perhaps a town named in his honor was Frederick's reward!

The heavy spruce tree forestation of the Salmon River delta, where Hyder was situated, must have provided ample reason for the pioneers to put their buildings on pilings rather than locate them directly on the delta. By the mid-twenties, there were more than 40 buildings joined by plank roads high above the tide flats. Like Hyder, B.C., sewage was not a problem with two tides a day to wash away the debris. The head of the Portland Canal has tides up to 26 feet high so the pilings had to be quite long. However, also like Hyder, B.C., fresh water was a problem because the well points brought up fairly salty water. Water bottles were filled at a creek 1000 yards away, but the majority just counted on beer and whisky for liquid refreshment!

When I was a boy in the thirties, my buddies and I liked to ride our bikes to Hyder to travel the plank roads, which were so smooth compared to our rough gravel roads. By the end of the mining boom and the Depression, the town built on pilings had been abandoned and thanks to rot, heavy snows and fire, it had virtually disappeared by 1950. All that is left today are a few rotten pilings sticking out of the mud flats.

In 1892, Capt. D.D. Gaillard of the U.S. Army Engineers surveyed the Portland Canal and built four storehouses. One is located at Eagle Point, right at the international border, and is still in fair shape. Some say it is the oldest masonry building in Alaska.

◄ INTERNATIONAL AVENUE IN HYDER, LOOKING EAST. FIRST BUILDING ON RIGHT IS NOW THE GLACIER INN. NEXT ONE WAS FORMERLY THE BANK OF HYDER, AND NOW THE FIRST CHANCE, LAST CHANCE SALOON.

ENTERING HYDER IN 1999 ►

Present day Hyder is located on the Salmon River delta, which was gradually cleared of the big spruce. The present population is about 100 compared to its highest peak of 500 in the heyday of the 1920s and its lowest of 37 people in 1917. Because access to the rich Canadian mines required travelling through Hyder and the Salmon valley, quite a benefit flowed to Alaska. An Alaskan friend mused once that the boundary survey team of 1903 must have included a few Canadian geologists who ensured that all the rich mines were on Canadian soil!

Although there have been many prospectors on the Alaskan side, the only producing mine was the Riverside, seven miles beyond Hyder. With a 100-ton mill it only processed 30,000 tons yielding 2,400 ounces of gold and 86,000 ounces of silver. But there was a fair amount of scheelite, the ore of tungsten, and in 1942, the U.S. government subsidized the owner to open the mine and produce tungsten for the war effort. That was on a small scale, though, and lasted only until 1945.

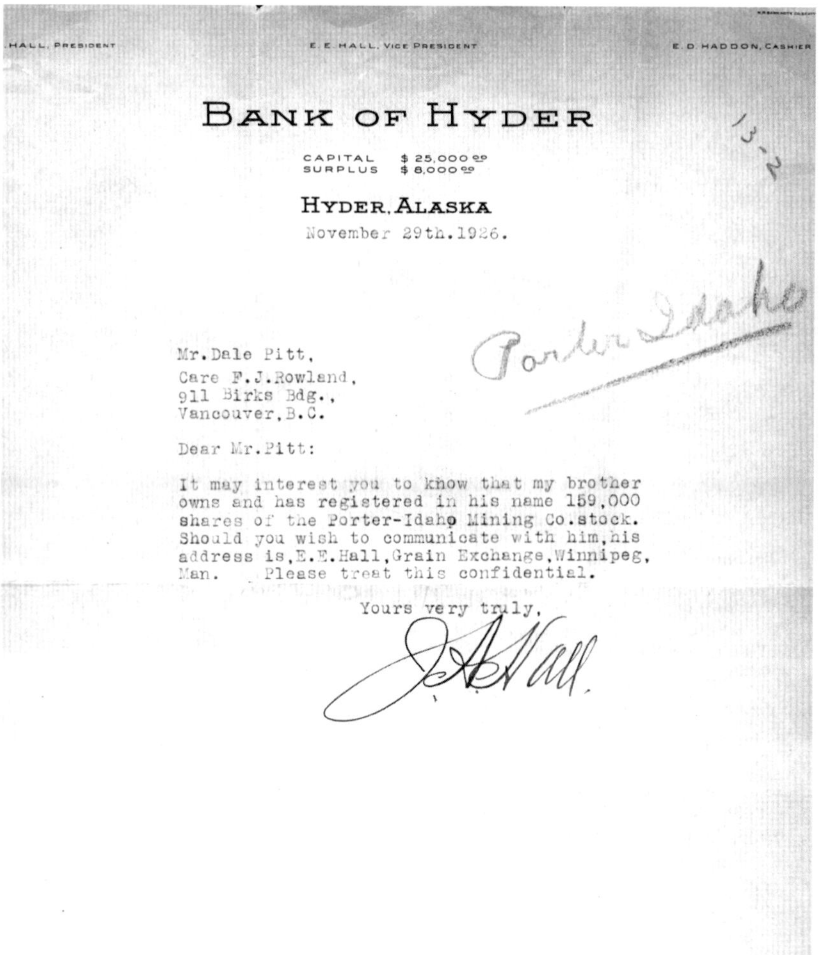

HALL, PRESIDENT E. E. HALL, VICE PRESIDENT E. D. HADDON, CASHIER

BANK OF HYDER

CAPITAL $ 25,000.00
SURPLUS $ 8,000.00

HYDER, ALASKA

November 29th.1926.

13-2

Porter Idaho

Mr. Dale Pitt,
Care F.J.Rowland,
911 Birks Bdg.,
Vancouver, B.C.

Dear Mr. Pitt:

It may interest you to know that my brother owns and has registered in his name 159,000 shares of the Porter-Idaho Mining Co.stock. Should you wish to communicate with him, his address is, E.E.Hall, Grain Exchange, Winnipeg, Man. Please treat this confidential.

Yours very truly,

J.A. Hall.

The Chickamin glacier area, 17 miles north of Hyder, was quite mineralized so the U.S. government also built a road to it. This project involved construction of a large steel bridge at Nine Mile and a smaller wooden covered bridge on Texas Creek. The latter was known as the only covered bridge in Alaska. The steel bridge was washed away when Summit Lake dumped its contents into the Salmon River in 1961.

Frank Blasher built a lovely two-storey log chalet at Chickamin in the twenties and stocked several lakes with trout. The fish thrived and local people enjoyed driving up to the beautiful Chickamin area for recreation. Now it is inaccessible except by helicopter but apparently the trout are plentiful. See Chapter 16.

The Bank of Hyder was the only bank in town and it was considered a wealthy institution during most of its life with capital assets of $25,000 during its heyday. The bank operated from the early 1920s until it closed in 1933 during the Great Depression.

◄ LETTERHEAD OF THE BANK OF HYDER. NOTE SMALL AMOUNT OF CAPITAL COMPARED TO PRESENT-DAY. THIS LETTER IS AN ILLUSTRATION OF PREMIER'S ATTEMPT TO TAKE OVER CLAY PORTER'S CLAIM TO THE PORTER-IDAHO MINE.

There must have been a lot of anti-union sentiment in Hyder in 1920 because Jack Sutherland of the Alaska Labour Union, who was trying to organize the workers of the day, was tarred and feathered in public, escorted to the weekly ship, and told not to return. There was no other union activity in the area until 1943, when the Mine Mill Union organized the Premier workers. It was about time as wages and conditions were terrible. In 1943 wages were $5 a day and the miners had to pay $1 a day for room and board. Today a good miner makes at least $250 a day.

A prominent Spokane businessman, Conrad Wolfle, was charged with mail fraud sometime during the 1930s in connection with his Cantu mining property at the toe of the Salmon Glacier. A very strange twist in the trial was evidence pre-

▲ THE COVERED BRIDGE THAT CROSSED TEXAS CREEK ON THE ROAD TO CHICAMIN GLACIER. IT IS NOW CAVED IN.

▼ THE RIVERSIDE MINE SEVEN MILES NORTH OF HYDER, 1940.

MV CHACON AT THE HYDER DOCK WITH MOUNT RAINEY IN THE BACKGROUND. THIS WAS THE WEEKLY MAILBOAT FROM KETCHIKAN. THE WATER AT THE NORTH END OF PORTLAND CANAL IS COMPARATIVELY FRESH SO WITH COLD WEATHER AND A NORTH WIND, THE BOATS COULD ICE UP.

sented of Wolfle's poetry, chief among them this verse that he used to advertise his mining stock: "Simple Simon went a-fishing in his mother's pail; "He feared the mighty ocean, so he never caught a whale. "To catch big fish - a little risk — you CANtu!"

Tommy Thomas and his wife May were the last of the Hyder old-timers and after 56 years of residence in Hyder, they left in 1977 to retire in the "Lower 48". Tommy was a real character who operated a gift shop right at the boundary. From the earliest days, the Canadian steamships carried tourists on the round trip from Vancouver to Stewart. They were always attracted by Alaska and would head there from the ships, walking the half-mile from the Stewart dock. Tommy was always there to greet them and sell them his souvenirs and trinkets. They would then proceed to the saloons where the bravest among them would get "Hyderized." This procedure involved downing, in one gulp, an ounce shot of pure grain alcohol (referred to as 'Snakebite') — if they were successful, they received a certificate of Hyderization.

I went to school in the same grade as the Thomas' son Burt, who was very tall. In Grade Eight we had a five foot female teacher and one day Burt was about to "get the strap" for some bad behaviour. When he stood up and put his hands out, the teacher couldn't reach them so she ordered him to take his punishment sitting down!

For many years, the Hyder children attended school in Stewart, while electricity, phone and medical services were also supplied from Canada. There was complete harmony between the two communities.

The longest serving resident of Hyder today is my friend Gary Benedict. He built the Sealaska Inn in 1964 and still operates it. Gary is a generally happy, jolly fellow who has had his share of escapades, including being shot. For years, his bar has celebrated July 4 with a wet T-shirt contest, which is one of the popular affairs in the north. In the busy 1970s, Gary laid out the low tide golf course on the tide flats alongside his premises and staged the "Ina Peterson Open." Ina was a well-liked colorful customer.

INA PETERSON, KNOWN AS HYDER INA. A WELL KNOWN LOCAL CHARACTER WITH A GREAT SENSE OF HUMOUR.

FAR LEFT: FRANKIE HALL, AN OLD YUKON '98 DANCE HALL GIRL WHO OPERATED THE ALASKA SALOON IN HYDER.

WHEN TELEGRAMS WERE USED, THEY WERE SENT IN CODE AS IT WAS TOO EASY FOR OPERATORS AND OTHERS TO PASS ON CONFIDENTIAL INFORMATION. THESE SHOW THE CODE MESSAGE AT THE TOP AND THE DECIPHERING AT THE BOTTOM. THESE MESSAGES ARE ABOUT THE PORTER IDAHO MINE AND THE PREMIER COMPANY'S ATTEMPT TO TAKE CONTROL OF IT, WHICH THEY WERE ABLE TO DO. H.A. GUESS WAS THE VICE PRESIDENT FOR THE GUGGENHEIM'S IN NEW YORK. DALE PITT AND BERT SMITH WERE THE LOCAL MINE MANAGERS.

RADIOGRAM

HYDER RADIO & TELEPHONE CO., HYDER, ALASKA

Ac'pt'd No. ...... Time Filed ...... M. Hyder, Alaska ...... Date Sept. 21st., 192 7.

Prefix ...... Sent No. ...... Sent to ...... Time sent ...... M ...... Check ...... Opr. ......

THE HYDER RADIO AND TELEPHONE CO. reserves the right to refuse to accept or transmit any message.

I agree that neither the HYDER RADIO AND TELEPHONE CO. accepting this message nor any other person, firm or Government concerned in the forwarding of this message shall be liable beyond the amount paid for sending same for any loss, injury or damage from non-transmission or non-delivery or neglect in relation to this message, or delay or error or omission in the transmission or delivery thereof through whatever cause such transmission, non-delivery, neglect, delay, error or omission shall have occurred.

To H.A.Guess,

Address 120 Broadway, New York, N.Y.

GREEN WUKIVXPILI VLOLZIQOQU VZYASEZHOP UMAUYUMUYV EZJAZUMBJA UMTAMIRAGD UOXOLEHTKU WGIPF UMEQO.

Porter Idaho Lowest Level struck ore at depth 400 ft. stop. Footwall side 2" assays about .02 oz. gold, assays about 14.8 Oz. silver 5" Assays about .18 Oz gold, assays about 649 Oz. silver, 3' side of Hang Wall assays about 2.8 Oz. silver.

B.F.Smith.

OUR CHARGE .		
OTHER LINE . .		
TOTAL		

RADIOGRAM

HYDER RADIO & TELEPHONE CO., HYDER, ALASKA

Ac'pt'd No. ...... Time Filed ...... M. Hyder, Alaska Premier, B.C. Date Oct. 12th 192 7.

Prefix ...... Sent No. ...... Sent to ...... Time sent ...... M ...... Check Nl. ...... Opr. ......

THE HYDER RADIO AND TELEPHONE CO. reserves the right to refuse to accept or transmit any message.

I agree that neither the HYDER RADIO AND TELEPHONE CO. accepting this message nor any other person, firm or Government concerned in the forwarding of this message shall be liable beyond the amount paid for sending same for any loss, injury or damage from non-transmission or non-delivery or neglect in relation to this message, or delay or error or omission in the transmission or delivery thereof through whatever cause such transmission, non-delivery, neglect, delay, error or omission shall have occurred.

To H.A.Guess,

Address American Smelting & Refining Co., 120 Broadway, New York City.

E. HALL IVAOO TVODE EBOYM ABENX ISZOZ OFIKE ZOEWE ESALW CANVC DAKQE BIQKO EK AKRBI BUHOB OQSCY OFFYK AKSSE DAOAK KALFU EBODA FUIBV SPLKC MEQEN DAOBO AKNKG KWUWC BJAXD AETOA HINLO TRYEM UOFAC FUICW AVRQI AUCEK IWDJU EWIFF UARGO SBIUU ESUVP BUHIU AHOIC AEEEP CIIGF SAPIZ BUHOB REGUN MIYND AHOIC IBGYM BJAXD.

E.Hall of Winnipeg has about 260000 shares Porter-Idaho which cost about twelve cents. Without committing Premier Gold Mining Company I asked, "Will you consi[der] selling for cash your shares at about twenty-one cents. He not part[ic]ularly enthusiastic but slightly interested. Then I suggested twent[y] five cents as a possible figure. Advised not sure whether you will [be] interested in buying but if he will make firm offer I would telegra[ph] you. He will consider and will advise in few days. Advise by teleg[raph] will you consider such purchase and maximum figure.

D.L.Pitt

OUR CHARGE .		
OTHER LINE . .		
TOTAL		

ADS FROM THE "NORTHERN ARGONAUT" NEWSPAPER, 1938. THE HYDER LUMBER AD SHOWS THE FAITH IN THE FUTURE THE OLD TIMERS HAD.

Salmon fisheries in Alaska have produced over 850 million dollars.
Gold production in Alaska is over 500 million dollars.

●

Our confidence in the North is based on

MINING
FISHING
AGRICULTURE
LUMBERING

and the

TOURIST INDUSTRIES

Build the Highway to Alaska and Watch our population Grow!

Hyder Lumber Company

R. L. Stewart — Roy McLeod

HYDER, ALASKA

Meat Market, complete line of Groceries, Gent's Furnishings, Hardware, Liquors and Gas

"If we haven't got what you want . . .
ask for it and we will get it for you."

EDGAR PROVINSE

General Merchandise

HYDER, ALASKA

Chapter Nine

HYDER, B.C.

HYDER, B.C., 1928. HYDER, ALASKA IS IN THE DISTANCE.

Hyder, B.C., was built adjacent to Hyder, Alaska in the early 1920s in response to the success of the Premier mine. With Stewart just two miles north and the wharf right beside them, the two Hyders had access to hundreds of miners coming and going.

Like its namesake on the American side, Hyder in British Columbia was built on pilings driven into the mud flats and adjoining the highway. It consisted of about a dozen buildings, including several hotels, a restaurant, the Canadian Customs, two whore houses, and three or four "blind pigs" — which was what unlicensed pubs were called in those days! Sewer service was not a problem as the outhouses were built over the tide flats too and twice a day, the tide came in and washed all the sewage away. Fresh water was a challenge, though, so the occupants drove well points into the mud and underlying gravel, and pumped up water that was too salty for drinking but adequate for washing and cooking. Besides, drinking water was not in great demand considering the vast amounts of beer and whiskey available.

The most famous resident of Hyder, B.C., was Bill Reid who became one of Canada's foremost artists. His father, Bill Sr., was a Scots American who owned the Keith Hotel, and his mother was a Haida from the Queen Charlotte Islands. Billy Reid and his sister Peggy went to school in Stewart and walked two miles between home and school daily. The Stewart News wrote in 1932: "Billy Reid passed to Grade Eight, second in his class of seven pupils and with sister Peggy left on the steamer for their new home in Victoria."

(Reid is considered by many to be the most famous and influential Northwest Coast artist in our time. His best-known creations are likely the five-ton "The Black Canoe" (i.e. "The Spirit of Haida Gwaii") at the Canadian Embassy in Washington, D.C., and "The Jade Canoe" at the Vancouver International Airport.) Bill Sr died at Stewart in 1942.

By 1946, both Hyders were getting pretty rickety. Two of the houses were lowered onto log rafts and floated to Stewart where they still stand. The rest deteriorated and collapsed, and by the mid-1950s the only evidence of the little towns were a few rotten pilings.

THE STAGE LEAVING HYDER, B.C. FOR PREMIER MINES. BUILDIING BEHIND LATER BECAME CANADA CUSTOMS OFFICE.

Chapter Ten

SCOTTIE GOLD

SCOTTIE GOLD CAMP, 1988

When brother Don and I started working for Morris Summit mines as labourers in February 1946, little did we imagine that 30 years later, Don would bring the property into full production through his position as president of Scottie Gold Mines.

Our first job on the property was to work on the Cat trains that were freighting supplies across the Salmon Glacier to Summit Lake, where the property was located. The freight boss was Oly Wilson, a fine old Yukoner who really knew his business. The mine manager was Evan Harris, a wild Welshman!

One day, Harris told me I was being promoted to supervisor. As I stuttered in astonishment and pride, he quickly explained that there would be no raise in pay and my crew would be two pack horses! It would be my job to haul mail and supplies over a five-mile route from Nine Mile on the Premier road over a horse trail to the Cantu Camp at the toe of the Salmon Glacier. We did not have horse snowshoes (which I probably wouldn't have known how to handle at the time anyhow) so we had to break trail in five feet of snow, a process that took several days. We built a small barn at Nine Mile and the job proceeded relatively well for supervisor Ian and his crew, the mares Dolly and Mabel. But a couple of interesting incidents tested my expertise!

One day, the freight was larger and heavier than normal as it included various machine parts for the mine. Dolly was saddled with the heaviest load of more than three hundred pounds. About a half-mile down the trail, Dolly simply lay down in the snow and refused to move. When a pack horse is down, it has to be unloaded, then raised back to its feet and then reloaded. Which is what I did, only to have the horse travel one more half-mile before taking another break in the snow! I realized then that she was complaining about the weight of her load, so I left part of it beside the trail. Dolly happily completed that trip satisfied, I'm sure, that she had shown the supervisor who was boss!

The mine maintained an office in Stewart, with which they had radio contact. When the boss had orders for me, the office man either drove the nine miles to the barn or sent a message. I received an urgent message one day that Gunner Gunderson, the mine cook, was sick and would need transportation from Cantu Camp to the Premier Road. Gunner was a huge man, about six and a half feet tall and 350 pounds. They transported him from the cookhouse to Cantu on a stretcher, inside a Bombardier snowmobile van. The snowmobiles were built on a truck chassis and in the right conditions could carry ten men or a ton of supplies. When the caterpillars plowed and compacted a trail for the snowmobiles, they could travel up to 40 miles an hour. But in fresh snow they were useless, and our horse trail was too narrow for any machine.

I recognized the challenge before leaving Nine Mile, so I decided to bring Gunner back on

CATS BREAKING TRAIL ON THE GLACIER.

a toboggan. Only one horse came with me on the trip to Cantu. The camp there had a sheet metal roof so with my hammer I pulled the nails out of one eight-foot sheet. Then I cut a six-inch sapling about the width of the sheet and rolled the metal around it, securing it with haywire to make the front of my toboggan. Small saplings attached to both sides of the sheet ensured that Gunner would not roll off.

When the snowmobile arrived, myself and another two men loaded him on the toboggan. I went in the lead, holding the horse's halter. We had attached two ropes to the rear of the toboggan and the other two men were responsible for holding those ropes on the downhill stretches to prevent the toboggan from crashing into the heels of my horse. (When something hits a horse's heels, they instantly break into a gallop and although I was leading the horse, he could well have knocked me aside and given poor Gunner a pretty wild ride!) Fortunately, the ride was uneventful and Gunner was successfully transported back to town and medical assistance. I never saw him again. In 1955, Granduc blasted the Cantu trail so that vehicles could pass on it.

PACK HORSES DOLLY AND MABEL AT MOUNTAIN BOY CAMP

The Morris Summit mine had been staked in 1930 by Ted Morris, an Australian who joined the merchant navy in 1939 and was killed somewhere on the Atlantic Ocean. Both the Premier and the Big Missouri companies completed some exploration in that decade but due to the relatively inaccessible location, they both gave up their options.

Exploration in the mid-1940s, when I worked there, was promising but the property was closed when the promoters' money ran out. In 1958, the Granduc road was built just across from the property, ensuring far easier access than ever before, and so brother Don formed Scottie Gold Mines in 1978 to recommence exploration. The early results were encouraging and led to development of a 200-ton underground mill. The mine yielded 100,000 ounces of gold, worth $50 million over a production period of five years. Those were the years when interest rates were twenty per cent and higher and with a large bank loan and a downward spiral on the price of gold, it was decided to close the property in 1985.

It remains closed but positive changes to the price of gold could quickly change its status.

1930 MAP OF MINING CLAIMS IN THE PORTLAND CANAL AREA.

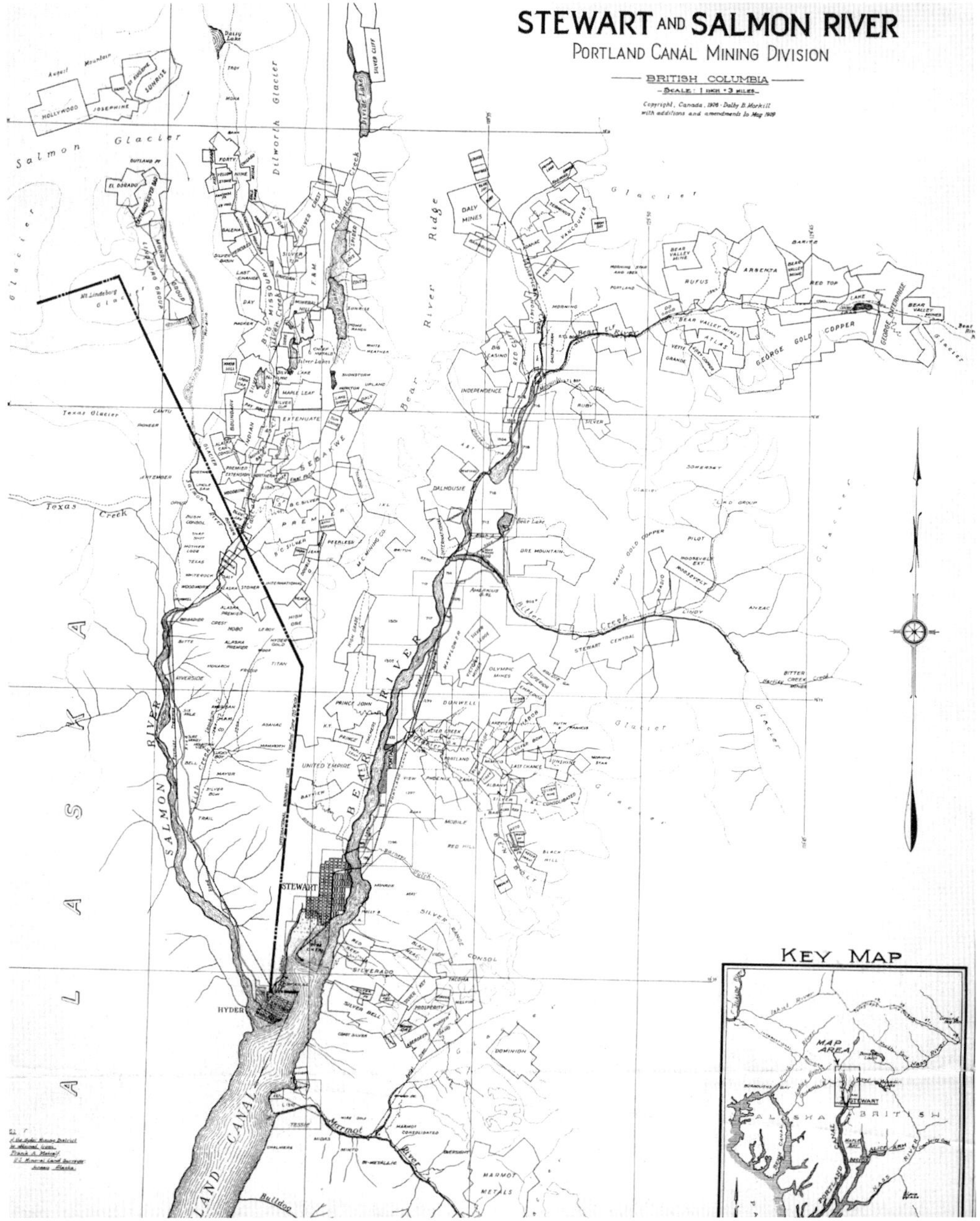

➤ DUNWELL MILL, 1929. THE CLEARED HILL SIDE BEHIND WAS FOR THE AERIAL TRAMLINE.

Chapter Eleven

DUNWELL MINE

The Dunwell property is located on the east slope of Bear Valley, four miles north of Stewart. It was staked by the Stewart brothers and Bill Noble. In 1925, they discovered a rich vein of gold and silver and commenced underground development. The ore was still going strong in the tunnel at 100 feet below the surface, and the promoters were excited. All the townspeople bought stock. It looked so good that construction began on a 100-ton mill. The rich ore on the surface was mined, sacked and delivered to the dock in Stewart with four horse sleighs.

The promoters planned a big celebration with a barbecue and free drinks. They also hired

➤ FOUR-HORSE TEAM IN 1937

a bagpiper to pipe the horse teams through Stewart's main street. The piper, Tommy MacLeod (no relation), was a veteran of World War I. As the entourage marched through town, my father and several other Scots were conspicuous due to their laughter, because only they, in the entire crowd, understood that Tommy was piping "The Barren Rocks of Aden."

The next level yielded virtually no ore; the vein had "pinched out" and the mill only lasted six months. It appears Wee Tommy was a better geologist than the promoters!

The mill building stood for many years until the big gravel avalanche of 1958 spewed forth millions of tons of gravel upon the site and damaged it severely. All evidence of the mill is now covered. The avalanche occurred at the same time as the Barney's Gulch cemetery was buried.

The All West Trading log yard is now located at the site of the Dunwell mill.

▲ DUNWELL ASSAY OFFICE AFTER THE GRAVEL SLIDE OF 1958.

➤ AFTER THE GRAVEL SLIDE OF 1958. NOTE THE SECOND GROWTH BEHIND.

Chapter Twelve

MURDER IN THE MINE

NEWHAWK MINE SITE. THE PORTAL IS AT THE FOOT OF THE MOUNTAIN.

In 1985, my brother Don launched a new corporation, Newhawk Gold Mines Ltd., and optioned the Brucejack Lake property in the Unuk River watershed. After a helicopter-supported diamond drilling program was successful it was decided to try some underground exploration.

The property is in a very isolated area surrounded by glaciers and large canyons but land access was finally established. That access involved using a barge across Bowser Lake, a road up the Bowser Valley to Knipple Lake, and then travel up the glacier to the mine site, a route of some 30 miles. (Knipple Lake was named after Charlie Knipple, a prospector.) When the freight route was established, a camp and equipment were moved in and work commenced.

The first priority was construction of a decline ramp at a fifteen per cent angle to be driven by rubber tire equipment. It eventually reached 700 vertical feet below the surface. From the ramp, tunnels drove off at intervals. When workings reach a certain depth, an underground dynamite storage area is permitted and it is referred to as "the magazine." The amount of dynamite permitted for storage is the equivalent of what is used in a single day.

In the case of Newhawk, storage permitted was 600 lbs and it was stored in a magazine some 600 feet down the ramp.

On Feb. 4, 1989, there were some 25 men working at the mine. Nine men were underground that day as well as two supervisors, Claude Webber and Otto Sawatsky. The supervisors had stopped to talk to a couple of workers about 200 feet away from the dynamite magazine, and one of these workers recalled later that one of the supervisors remarked: "I smell fuse burning. Let's go see what it is." Shortly after that there was a terrific explosion.

A dynamite blast creates a lot of smoke and noxious gas but miners carry a type of gas mask called a self-rescuer. The crew donned their self-rescuers and made their way to the surface, not knowing what the explosion was. After the smoke had cleared, mine manager John Michelle took some men and went back down the ramp to check. They found havoc, including the bodies of the two supervisors and a wrecked magazine. Apparently, the concussion from the explosion had been sufficient to kill the supervisors who were standing about 150 feet away.

Two diamond drillers, Ray Felado and Danny Melanese, were working in the deepest part of the mine, and they thought the blast was a normal part of the work cycle. When it got pretty smoky, they simply put on their self-rescuers and kept drilling until the manager found them and told them what had happened.

Upon investigation by the Department of Mines and the RCMP, it was determined that the blast was deliberately triggered. But why and by whom? The first suspect was a young man who had been chastised that morning but apparently a lie detector test cleared him. No other suspect ever surfaced, at least to my knowledge, and the killings still remain a mystery. Newhawk Company announced a $25,000 reward but not a single tip was ever received. No-one appears to have benefited and so a motive was never established. However, the RCMP have kept the file open as they do in all unsolved murder investigations.

In 1991, with a combination of low gold prices and few investors, it was decided to close the mine. After a few years, when conditions failed to improve, the property was mothballed and eventually sold. There is a half-million ounces of gold and about seven million ounces of silver yet to be exploited when prices improve.

After the closure, but before the camp was dismantled, I visited in early spring. As we approached by helicopter we noticed a lot of grizzly tracks in the snow. Upon landing, we could see that a bear had ripped the plywood off the back of the kitchen where canned food was stored. Scattered on the snow were several cans of pressurized oven cleaner cans, some of them punctured by bear teeth. From there, the bear tracks made a bee line straight up the hill behind the camp. I'm sure Mr. Bruin didn't care for that treat, and there is no chance for him to return for something tastier as the camp is now gone and the land returned to its native state.

DAILY DELIVERY OF EXPLOSIVES ON THE WAY TO STORAGE IN THE UNDERGROUND MAGAZINE.

Chapter Thirteen

NASS GOLD RUSH

PUMPING WATER FROM ALOFT

In 1910, a prospector arrived in Stewart with word of a big placer strike on the White River, a western tributary of the Nass River. A mine rush ensued, and the route chosen was up Bitter Creek to the Big (Bromley) Glacier, over it, and down into White River. Two dozen men joined the enterprise. At their destination, they found some fine gold in the black sand beside the river so commenced sinking a shaft because gold always sinks through the gravel thus making the richest strikes at bedrock.

Pumps (like the one in the photo) would have been built on site with lumber that was whipsawed from native trees and only the bolts, nails and nozzle packed in.

Whipsawing was a technique to make lumber by hand labour. A platform would be constructed to hold the log that was ready to be sawn. One man would stand on the top of the platform to pull the saw up with another man below to pull the saw downward. The process was helped by gravity but it was slow and tedious work.

The effort of sinking the shaft was a failure due to very large boulders close to the bedrock, as well as too much water for the pump to handle. So yet another gold rush was stymied and the prospectors had to start looking all over again. As they used to say: "The gold rush is over and the bum's rush is on!"

TRAVELLING ON THE GLACIER TOWARD THE GOLD STRIKE.

Chapter Fourteen

ESKAY CREEK – UNUK RIVER: The Richest Claim in British Columbia

ESKAY CREEK CAMP

In 1989, when Murray Pezim's Calpine Mining hit the bonanza at famous Hole 109 on Eskay Creek, that name became instantly renowned in the mining world. But I know many pioneers who had worked that area years earlier. In those days, we called it the Unuk, which was the name of the river that drained into Burroughs Bay on the Alaska side.

Tom Mackay and his associates were the first stakers of what is now the Eskay Creek mine. In 1932, using a float aircraft, they landed on Tom Mackay Lake and after prospecting the area, staked the location of the present-day mine. Tommy was the assistant manager of the Premier mine; he had started there as a dishwasher while paying his way through university.

His associates were all supervisors at Premier and while not one of these associates profited a nickel from the property, their names have been attached to landmarks in the area. Harry Melville, a mine surveyor, was immortalized in Harry Mel Creek; Bill Prout, an engineer, in Prout Plateau; and Sam Coulter, the master mechanic, in Sam Coulter Creek.

Although the property was diamond drilled several times with indifferent results, Tommy never despaired and held title to the property until it was optioned to Pezim. By the time the ore was hit, Tommy had passed on but his wife became very rich thanks to his foresight.

Tommy pronounced his last name the traditional Scots way as "makii." Besides the prospecting, he enjoyed a very successful career. When Dale Pitt, the Premier manager, was transferred to Australia to operate the Big Bell mine there, Tommy went along as assistant manager. In 1940, Tommy joined the Royal Canadian Air Force and by the war's end, had been promoted to Wing Commander T.S. Mackay MBE, chief intelligence officer, in an overseas role.

Upon leaving the RCAF, he returned to the Portland Canal district and his dreams for the Eskay/Unuk property. He raised some money, hired a hand-mining crew in Stewart and started organizing for work in the summer of 1946. The tunnel he was going to drive was some miles from the lake where the aircraft would land, presenting the challenge of transporting supplies. Over a drink with pilot Pat Carey and Bill Crawford, the pack horse train owner, he asked their advice. Bill's solution was to use a gentle little pack horse named Tony, who only weighed a thousand pounds, and Bill said he knew how to get Tony ready for an airplane ride if Pat would only consent to transport him. Cocky as ever, Pat declared it would be no problem. The Norseman aircraft that Pat flew could handle two thousand pounds of freight.

A few young fellows, myself included, were recruited to help. Bill first tied Tony's legs together and we gently tipped him onto his side into a rope cargo net. The doors had been removed from the Norseman and a ramp installed so we dragged the cargo net, with its precious cargo, up the ramp and into the aircraft. We tied the load down and Bill sat on Tony's neck with a big hammer in his hand, just in case

PAT CAREY, WHO FLEW TONY THE HORSE TO TOM MACKAY LAKE IN 1946, IS PICTURED HERE WITH A WHEEL-SKI BEAVER AIRCRAFT ON TIDE LAKE FLATS.

he needed to administer a quick sedative, and away they flew. Pat told me later that the horse behaved perfectly. When they landed at Mackay Lake, pulled Tony off the plane and untied him, he just stood up, shook himself a little and started eating grass.

Tony was put into harness and, pulling a go-devil, he hauled all the freight to the mine site. He also became quite a pet to the crew.

The hand miners, all in their late fifties, included Tony Mariotti, Herbert Anderson, Eric the Bear and Pontus Johnson. They all had silicosis, but nonetheless worked into December until all the dynamite had been used. (Silicosis is a lung disease caused by acid formed by silica dust and was and is irreversible. All the old miners came down with it due to the accepted procedures of dry drilling and entering the work place immediately after a blast.)

By the time I started in the big mines water was used extensively in drilling and wetting down the ore after a blast. But the experts of the day decided aluminum dust would neutralize the silica and that dust was blown around our change rooms. Later it was decided that aluminum was not helping but a few of my old time workmates with Alzheimer's disease are blaming the aluminum for their present condition.)

No aircraft with skis was available to fly them out so it was decided the miners would snowshoe the 40 miles to Burroughs Bay and take the boat back to Stewart. Fortunately, that tough Finn, John Lehto, was trapping in the area and volunteered to break trail for the miners. Mariotti told me later that they would never have made it without John.

When the miners arrived at salt water in Alaskan territory, they hired the Matneys to take them 40 miles by small boat to Ketchikan. The Matneys ran a ranch, but were always happy to pick up extra money by transporting mining people to Ketchikan.

From Ketchikan, Fred MacKay operated the weekly mail boat to Hyder so, after a few days of partying, the crew headed out with Fred. But apparently Anderson hadn't finished having a good time and was still sipping on a bottle of rum. He went on deck to relieve himself and when he didn't return, someone went to look. He was gone. So Fred turned the boat around and everyone came on deck as spotters. They soon found Herb, treading water and occasionally taking another drink from the bottle. By the time they retrieved Herb from the frigid water, a half-hour had elapsed and the bottle was almost empty. Fred wanted to return him to Ketchikan, which was closer, but Herb would have none of that — all he wanted was dry clothes and his other bottle of rum so they continued on uneventfully to Hyder and Stewart.

Tommy and Squeaky Alex McKay, the cook, stayed in camp as Alex was quite crippled and also had a silver plate in his head due to wounds from World War I. By Christmas, the snow was eight feet deep. The hay and oats for Tony the horse had run out, and the snow was too deep for him to forage. Tommy took the rifle and put one shot behind Tony's ear. He said later it was the toughest thing he ever had to do.

Tommy was quite a storyteller and the following one I heard from him a number of times.

As assistant manager at the Premier mine in the '30s, he did a fair amount of traveling. One time, he was in Toronto on business and staying at the Royal York Hotel, which was then, as now, a prominent hostelry. He signed the register "T.S. Mackay, Premier, BC." The name "Premier" was all the address ever needed because the postmaster always distributed mail personally. Tommie put his bag in the room and went out to do his business. When he returned that evening, he discovered a bottle of champagne on ice and a bouquet of flowers on the table. The card read: "To the Premier of British Columbia, with best wishes from management and staff of the Royal York Hotel." Tommy hesitated for only a moment. Then he decided that to correct the mistake would only embarrass the management, so he uncorked the bottle and quenched his thirst after a long and tiring day!

Tommy McQuillan (with namesake McQuillan Ridge) was the real pioneer of the Unuk River

PROSPECTORS TENT ABOVE TIMBER LINE

area and also the prospector responsible for the staking of the Granduc mine. He was not only a friend, but also one of the finest men I have ever met. McQuillan prospected the whole area and spent many winters on his trap line there.

Mackay and McQuillan joined forces on the Eskay property in about 1980, naming their property Stikine Silver Mines, which eventually was the entity to option claims to Pezim's Calpine Resources. After Hole 109 was drilled, the stock in both companies skyrocketed and, like Mackay, McQuillan also left behind a rich widow.

We called them "the two Tommies."

Two other pioneers in the Unuk watershed in the 1930s were the Johnstone brothers, Bruce and Jack (with BruceJack Lake named after them), who hailed from Ketchikan. They had a riverboat with an outboard motor and would use it to get as far upriver as conditions allowed.

Rumors circulated at one point about a terrible groaning that was occasionally heard by backcountry travelers. The Johnstones deduced that it must be a big mangy grizzly bear that was making the sounds. Over the years, the bear became known as "Old Groaner." Grizzlies didn't have many friends in the area, because of the damage they would wreak on the cabins and caches. In 1936, Bruce spotted a bear on a river bar and shot it from the boat. When they got to the carcass they were amazed at the condition of the head. It was badly scarred, one eye was gone, and the jaw joint on one side appeared to be very badly damaged. They removed the head and took it into a doctor in Ketchikan. His autopsy found two .33 slugs, one beside the eye and another at the jaw joint. Both wounds had hit bone and the bone had calcified around the wound, undoubtedly causing great pain and the reason for the groaning. But the doc wasn't finished; he cut into the cartilage at the front of the neck, retrieving three .38 revolver slugs all indicating that someone lying on the ground under the bear had fired them.

Connections were made: in 1929, trapper Jess Setherington had left Stewart to trap at the Unuk, never to be heard from again. He had with him a .33 calibre rifle and a .38 pistol.

I believe the bear skull is still in the Ketchikan museum.

Once Pezim's drilling outlined the rich ore body at Eskay, a road was built to join the Stewart-Cassiar highway at Bobquinn Lake (Bob Quinn was a line man on the Yukon Telegraph in this area) and production commenced there in 1994. The ore was so rich, it was shipped raw. Half went to the port of Stewart for shipment to a smelter in Japan and the other half went to Kitwanga, B.C. from where it was transported to a smelter in Quebec. Currently, in 2003, the Eskay Creek gold and silver mine is one of the richest in the country. Production at the end of 2002 was 2.3 million ounces of gold and 100 million ounces of silver. Reserves were about 1.8 million ounces of gold and 80 million ounces of silver so Eskay Creek will undoubtedly be the richest gold and silver mine in the history of British Columbia.

I was always amazed by the enduring faith of the two Tommies in that property, and I'm sure if they are looking down they are feeling mighty pleased. But Pezim and his geologists also deserve much praise for their efforts. It is amazing that they would diamond drill 108 holes with mediocre results, and not give up. A good mine is like a good woman — hard to find but definitely worth the search!

STAKING CLAIMS ON MOUNTAIN TOP
UNUK RIVER VALLEY

Chapter Fifteen

THE EAST GOLD GROUP

EAST GOLD CABIN IN 1947. AS A MEASURE OF THE SNOW BANK, CONSIDER THAT THE ROOF PEAK IS TWELVE FEET HIGH.

When Tide Lake drained for the last time in 1931, prospectors staked the perimeter that had previously been under water. They discovered large zones of minerals on the west side and interested the C.M. & S. Company in the showings. The company did some diamond drilling without results except that one hole hit an extremely rich gold and silver vein about one foot wide.

A big company doesn't fool with such small widths, so they let the claims revert to the owner who was Mrs. Campbell. She had no problem leasing the property to some competent miners who drove a tunnel and found the ore body at the diamond drill intersection. The gold and silver was found as electrum, which is a natural amalgam of the two metals. Some was fabulously rich, running $100 a pound after sorting, which meant a man could carry out 50 pounds to town and get $5000 for it in even during the Hungry Thirties.

But the pockets were small and erratic. The miners would hit a rich spot, then spend its value tunneling for the next pocket.

When Mrs. Campbell died in 1940, her niece Peggy Philips and her husband Al inherited the property. As Al was an experienced miner they started work on the claim again. In January 1946, Al hired my brother Don to work with him and "Three Finger" Bill Riddell. Peggy was the cook.

At the time, I was working with Arthur Cameron on the Marmot River. In February, a messenger came to tell me that Dad wanted me in town right away. When I got there, my father told me that Don had a poisoned groin and that Al and Three Finger had snowshoed out for help, leaving Peggy to care for Don, who couldn't walk. Fortunately, two native trappers, George Brown and Tommy Danes, were in town with their dog teams. George, Tommy, my pal John Haahti Jr. and I geared up and headed out immediately to Big Missouri where we spent the night. Early the next morning, we headed to Long Lake, travelled over Betty Glacier and went on to Tide Lake. The journey was terrible with a great deal of fresh snow so that the snowshoes would sink up to our knees and the sides of the tracks would cave onto the snowshoes making every step very heavy. As well, it was white out conditions, which means that there are no shadows and you can't identify the terrain — we couldn't tell if we were travelling on a side hill or a flat. I tumbled many times into little gulleys as I was breaking trail because I couldn't see them. We finally arrived at Tide Lake flats in the dark and although we were worried about crossing the Bowser River, we managed without incident.

Prior to Tide Lake 'dumping,' all the timber was above the shoreline and that's where the cabin was located so we knew we had to climb 500 feet but, in the darkness, who knew where? Thanks to Peggy, we soon discovered a guide — it was a naphtha light that she had hung in a nearby tree.

Soon we were safe in the cabin, exhausted but delighted with the grub that Peggy had ready. The next day we were in no shape for the return trip so we rested, which was fortunate because that night a high wind created a wind crust on the snow. It made our transportation of Don relatively easy the following day as the dog sleigh rode well on the crust. We still had to wear our snowshoes but we walked right on top so the walk out was a breeze compared to the walk in.

We had easy going up the glacier and went on to Premier mine the same day. Premier's Dr. Joe Payne later told me we were just in time and that a few more days would have been too late for Don. It was the first close shave for brother Don but it wouldn't be his last.

Al Phillips continued mining at East Gold, finding high-grade ore and then spending what he made looking for the next pocket — just like the miners before him. One day in 1968, Al stayed too long after he lit the fuse and the explosion killed him. The property has been idle since.

EWIE LANGILLE CROSSES THE SMALL LAKE AT THE TOE OF THE BERENDEN GLACIER, 1938. DOWNSTREAM THE BOWSER RIVER WAS TOO HIGH TO WADE AND TOO FAST TO RAFT, SO THE PROSPECTORS USED THIS METHOD TO CROSS

Chapter Sixteen

HIGH GRADE

THE CHICKAMIN GLACIER IN 1935

One of the smallest properties with the highest grade of ore in the Portland district was the Larry Thornton prospect at the Chickamin Glacier on the Alaska side. It was also a very unique operation in the thirties.

Larry had found some rich float (loose rock) at the toe of a small glacier at the 5000-foot elevation. But it seemed pretty positive that the source of the float was under the glacier so Larry's solution was to mine the ice!

As the glacier was well above the timberline, he had to construct a cabin for shelter. Next he acquired a small steam boiler fired by oil. Then he ran a hose from the boiler to a six-foot piece of pipe that blasted out enough steam to 'drill' the holes into the ice — by melting it! When Larry had drilled a sufficient number of holes, he blasted and hauled the ice out of the tunnel by

➤
LEFT TO RIGHT:
JACK McLEOD, GEORGE KUSTUS, CHARLIE LUNDSTRUM
IN THE HEATHER AT CHICKAMIN GLACIER AREA

wheelbarrow. In one season, he advanced the ice tunnel 100 feet and found the source of the float. Unfortunately the vein was very narrow and short so, although it was very rich, there was not enough value to warrant the cost of production. So poor Larry had to give up and go looking for something better!

* * *

The Mountain Boy Group on lower American Creek produced some of the richest silver ore mined the district. As early as 1910, ten tons of ore, running 1200 ounces to the ton, was shipped to the smelter. In 1935, miners discovered a rich ledge on a very steep cliff. Using a system of wires, they managed to get another ten tons down the cliff to the pack horse trail but, as usual for the area, it cost more to get the ore out than the smelter would pay. In the late 1990s, my pal and school chum, Rudy Kasum, took an option on the property. He first built four miles of road from the highway to the mine and instituted a program of diamond drilling but the results were not promising. Unfortunately, Rudy passed away soon after but I believe his sons are still working the property.

CHICKAMIN CABIN BUILT 1920 BY FRANK BLASHER ➤

Local Characters

Chapter Seventeen

DAGO MARIE

MARIE WITH THE STAKING CREW, RETURNING FROM THE NEW MANHATTAN PROJECT. FROM LEFT, ANGELO BUGNELLA, RUDY KASUM, MARIE AND GEORGE LUPICK.

Premier Mine was located in British Columbia territory but it could only be accessed from Stewart by travelling through Alaska. The road entered Alaska about a half-mile past the Stewart dock, wound on for about thirteen miles, and then came back into Canada. Premier, which was about three miles further down the road, was a company town with management that did not permit alcohol or related rowdy activities!

Enter Marie and Paul Meger in the early '20s. They built two houses right at that international boundary, only about 100 feet apart; one was located on the Canadian side, and the other on the Alaska side. Although minerals had never been discovered there, the Megers staked claims on the property underneath both buildings in order to secure title — and the end result was indeed a little gold mine!

Marie and Paul had a neighbor named Little Joe Frank. Joe, an Italian immigrant and placer prospector was only five foot two but he was a going concern. His shack was built right on the International boundary right of way, no man's land, and as far as I know none of the authorities ever bothered him. Joe had a trail down the hill to the Salmon River where he panned for gold. He did find some as he would show us his dust and small nuggets. In the late forties he retired and moved to Stewart where he bought the Herb Hunter shack at Barneys Gulch and spent his last years. After Joe passed away the word was out that his gold was unaccounted for and for a few years his property was searched with metal dectectors and some holes dug here and there. But I never heard if anyone found his stash. I wish I had a better lost mine story for this book so we could have one more gold rush. Like the prospectors prayer, "Dear Lord, please give me one more gold rush and this time I promise not to pee it all away."

Half a mile downstream from the Meger's place is a corner at the top of the hill beside the big gravel pit. That corner is still known as Minnie Jones'. Minnie, who was a coloured lady, had a bawdry house there in the twenties and I do remember the remains of a house at that location when I was a boy in the thirties.

The Meger enterprise, called Thirteen Mile, quickly became well known. The building on the Canadian side was developed as a house of ill repute. Marie, the Madam, hired several girls whose work was completely legal according to Canadian laws of the day. Those same lawmakers, however, were very strict about unlicensed liquor. But the Americans didn't mind bootleggers (especially if the customers were Canadian miners) so Paul's house on the American side became the watering hole. With 200 single miners housed at Premier, only three miles away, the trail from the mine to Thirteen Mile — and between the two houses — became extremely well worn. One Premier miner, Jimmy Wood, liked the set-up so well he never got past Thirteen Mile for sixteen years!

In 1940, Canadian authorities closed all whorehouses to prevent disease during World War II and that was the end of Thirteen Mile.

Marie, who was affectionately nicknamed Dago Marie thanks to her Italian heritage, had been born in the Manhattan district of New York City and she had a delightful accent. She never drank alcohol, leaving that up to her husband, who cheerfully drank enough for both of them. Marie's vice was gambling with Canadian penny mining stocks, and she would read The Northern Miner from front to back. She was determined to make her fortune from mining but stocks on the

Vancouver Stock Exchange never seemed to grow — any profits were quickly ploughed back into the ground by way of the Howe Street promoters.

Marie and Paul left the Stewart area for ten years but returned in 1950, purchasing a building in Hyder that is now well known as the Glacier Inn. (Over the course of a half-century, the walls in the Glacier Inn have become papered with dollar bills, on loan from customers who autographed the bills in case they should ever return in desperate need of money for a drink!)

The Megers ran the Glacier Inn for about 10 years. Marie didn't have a till for her money, and she didn't like leaving cash lying around. So it all went into her ample brassiere. The big bills were tucked away on her right side, which the miners called "The Bank of America" while the smaller bills snapped into the left cup, whose name was the Bank of Montreal — which was the only real financial institution in Stewart at the time.

During the decade away from Stewart, Marie's mining fever did not subside. In 1954, my brother Don and I thought we should stake some claims adjacent to the Granduc project, which was just getting underway. Marie overheard us discussing plans and insisted that she be allowed to help finance the venture. She quickly became the financial partner and we called ourselves the Manhattan Syndicate in her honour! So we lined up a tractor with an Athey wagon, including tent and stove, and headed up to the Salmon Glacier and then over to Leduc in mid-winter to stake a bunch of claims. We didn't really know what we had staked, due to the snow covering, but felt we couldn't wait until spring.

We had hoped to sell the property as moose pasture to some promoter but no luck there. However, one day my phone rang and it was Marie, who was very excited.

"Ian, Ian, I just talked to Larry Postle and Jack Crowhurst (Granduc management men). They were in the bar for a drink and I told Mr. Postle that I knew he had a good mine at Granduc but we have the Manhattan, the mother lode. And Larry said, 'Yes, Marie, you have the monzonite!'"

His comment must have sounded impressive to poor Marie, but I knew that monzonite was only a form of granite so I had to tell her: "Yes, Marie, we have a mine that will produce tombstones but I'm afraid there's no market for them here."

In 1960, Paul and Marie sold the Glacier to Lloyd Filion and retired to Portland, Oregon where they both passed away a few years later. The Glacier Inn still belongs to the Filion family.

Chapter Eighteen

DAVID GUNANOOT AT OUR UPPER AMERICAN CREEK CLAIMS, 1958

DAVID GUNANOOT

David Gunanoot, a Gitsan native, was born the year before his father, Simon Peter, was charged with two counts of murder.

The charges were laid in Hazelton, B.C., in 1906, and Simon Peter fled the area with his wife and children into the wilds of northern British Columbia.* With a reward on his head, he eluded the law until 1919 when he gave himself up and subsequently was found not guilty.

Simon Peter's favorite wilderness headquarters was Graveyard Point at Bowser Lake, 30 miles north of Stewart. After being acquitted, Simon Peter and the family trapped in the Bowser and the Bell-Irving River watersheds, both of which were relatively inaccessible. They came to Stewart to sell their furs and to purchase supplies, and it was during these visits when I came to know David.

But it wasn't until the early 1950s that David and I became close friends. We worked the pack trains together and went on several prospecting expeditions. David was an amazing man. He never attended school but taught himself to read a bit, and he was a highly skilled bushman. I'll never forget how he led me through thick forests for two days without even an axe blaze, unerringly guiding us to a copper showing he had seen 30 years earlier. He could also build snowshoes as perfect as I've ever seen.

Like many of his race, he also excelled in the oral tradition of storytelling. Here are some of David's stories.

"My first memories as a little boy were of traveling with a dog team through the bush. When we stopped at night, Father would erect a white man's type of camp tent and set up a B.C. heater inside the tent with several lengths of stove pipe to the outside so it was quite warm. (A B.C. heater was a sheet metal stove.) We sometimes travelled hundreds of miles in the winter up to Spatsizi and all over to Manson Creek where

A COTTONWOOD DUGOUT CANOE BUILT BY THE GUNANOOT FAMILY ON MEZIADEN LAKE, 1928

Dad panned for gold. When Dad told us stories about towns, where a lot of people lived, I thought of a whole bunch of tents all with B.C. heaters! But our main base was Graveyard Point at Bowser Lake and Dad felt very safe there as we could see in both directions for many miles. When Dad needed supplies, he would take his furs over to the Telegraph Trail and George Beirnes would sell them and bring back the supplies we needed. George was a very honest white man."

When David was 12, he remembered his dad saying: "I'm tired of hiding, I'm going to give myself up and these white men will hang me up." David said he wrapped his arms around his father's legs and said, "No, no, Daddy, I won't let the white mans hang you up."

Treaty Creek is a tributary of the Bell-Irving River and was named to commemorate the peace negotiated between the Stikines and the Gitsan who had a fight over rights to the valley. David showed me the gun pits that his people had dug at Owee Gee River for their defence. The Bell Irving River was named for D.H. Bell-Irving, the first Canadian officer killed in action during World War I.

Bob Switzer, formerly of the Canadian Broadcasting Corporation, is my good friend and we share a vacation property on Meziaden Lake. He became equally as fascinated with David's stories, and we spent countless hours around the fire, listening to them. David claimed that the original map surveyors gave the wrong native names to a number of landmarks. He said the Meziaden name should have been attached to what is now Bowser Lake. In the Gitsan language, Meziaden means "milky water" and, while the present Meziaden is crystal clear, the Bowser is glacier-fed, grey and murky. David said the proper name for Meziaden is "Sushiadit" meaning "where the fish spawn." The white man named the Bowser for the B.C. premier who was in office in 1915. I have a 1910 map of the area that shows a lake and river sketched in with dotted lines approximately where Tide lake is now, and the map refers to it as Brightwell Lake. See Page 1 Map.

I once asked David what he knew about the large burned out areas of forest in the Meziaden Lake – Stikine Creek district. He told me: "My people started the forest on fire when there was hot weather and a good wind. When the fire burned out, about two years later, the blueberry bushes would come up and we would harvest lots of berries. Also the willow and alder bush would start to grow and the moose would come into the country. Heavy timber no good for berries and moose." After 70 years, the second growth coniferous trees have choked out the sunshine for the berries, and soon there will be a shortage of moose feed again.

Simon Peter died at Todd Creek in 1933, en route to Stewart for some trading business. The family buried him at Graveyard Point beside his father, Nah-Gun. I passed by the Point in the 1950s and there was a nicely whitewashed fence around the graves. When I went back 30 years later, the site had returned to nature, leaving barely a mark to show that the spot had ever been occupied.

David's affectionate name for me was "Chief." When I was courting my future wife, Vivian, we would often spend the day hiking in the mountains around Stewart and it would be late when we returned to town. The restaurant would be closed but, not to worry, I owned the place and so we would head for kitchen, fire up the grill and cook up steak and potatoes. Once in a while, as we were tending the grill, I'd lean over and nuzzle Vivian on the cheek. One night, I thought I saw a shadow moving near the window but thought nothing of it. A day or two later, I met David on the street. He grinned at me and said softly: "How is the kissing chief today?" The moving shadow had been David investigating the light in the restaurant!

During World War II, David spent some of his trapping profits to purchase a house on Second Avenue. It had formerly been used as a house of ill repute, referred to as Louise's, until the government closed all such establishments in B.C. in 1941. It was David's home for the next 40 years; in very recent years it was moved to Stewart's main street and is now part of the "Inn on Ripley Creek" complex.

In the late 1950s, David worked as the bull cook at the Leduc camp where Granduc Mines was diamond drilling and exploring the ore body. The winter camp was located right on the glacier and consisted of a dozen huts plus cookhouse. Each unit had an oil heater with a small tank that had to be filled daily. One of David's chores was to refill the tanks: he would use a hand pump to draw the oil from the big 45-gallon drums into a smaller can and then pour it into the individual heaters. The empty drums were stored at the end of the airstrip until they could be flown into town and refilled.

Ski aircraft serviced the camp regularly, but the winter of the year in question was particularly vicious and the planes couldn't fly. Superintendent Stan Hunter recognized that, although the food on hand was sufficient, the supply of heating oil was dwindling quickly. He gave David the job of trying to eke out as much remaining oil as possible from the "empty" drums stacked near the airstrip. David was sure he had emptied the drums the first time around, but the boss being the boss, he set to work with his hand pump. After a long couple of days of trying to get blood out of the proverbial stone, David had only rescued a few gallons. "You might as well give it up," Stanley sighed. And David mildly replied: "Stanley, you should have told me before that you wanted some oil left in those drums."

In 1968, Pierre Berton authored a book entitled My Country, which included a story about Simon Peter and words to the effect that David had an extensive criminal record. I knew this to be untrue and when I told David about the book, he was disturbed and asked me if I could do something about it. I was mayor of Stewart at the time and the member of a coalition of northern communities that was fighting for reinstatement of a federal government subsidy. (The government had just cancelled the subsidy to Northland Navigation, which provided freight and passenger service on the north coast. We ultimately failed in our efforts.) But through that coalition, I had met an impressive young lawyer named

DAVID PREPARES FOR WINTER TRAPPING ON THE BELL IRVING RIVER IN 1968 WITH A NEW SNOWMOBILE AND NEW SNOWSHOES, CRAFTED BY HIMSELF.

Carey Linde from Queen Charlotte Island. I asked if he would consider representing David in a civil suit against Berton. He took the case on the basis that he would receive one-third of any settlement.

Linde started an action against Berton and the publishers, McLelland & Stewart. To Berton's credit, he recognized that a mistake had been made and he flew to Stewart for a meeting. The result was a $12,000 settlement for David, the biggest chunk of money he had ever seen. As his adviser, I explained some of his investment choices such as a savings account at the bank!

"No," he decided, "I want to buy a red pick-up truck." David was in his sixties and had never driven a motorized vehicle, but this had been his dream. And it came true — except the truck he bought was green!

David passed away in 1990. I was a pall-bearer at his funeral in "old town" Hazelton's Anglican Church and it was the most impressive memorial service that I ever attended.

It was a combination of old Church of England rites and native customs. Rev. Brian Wood, an old friend and neighbour of David's from his mining days in Stewart, conducted the service along with some Gitsan chiefs. David had been a chief — Chief Legap — and his cloak was folded beside the open coffin in the church hall. His big strong hands were folded below his breast. When the ceremonies started, the women filed past the coffin and wailed loudly. After several rounds, the coffin was closed, David's cape laid on top and we pallbearers carried him out of the hall and into the church. Brian started the service with prayers and hymns. Then the chiefs chanted words in Gitsan, removed the cape from the coffin, and draped it on David's nephew, Jerry, who became the next chief. Finally we took the coffin out to the hearse and rode up to the old cemetery where again the women filed past, wailing. Then Brian said the final words and a great old friend of mine was laid to rest. But never forgotten.

* Several books have been written about Simon Gunanoot's flight. My personal favorite is "Trapline Outlaw" by David Ricardo Williams (Sono Nis Press)

PREMIER CAMP FOUR, 1927.
BATTERY POWERED LOCOMOTIVE.
MINER USING CARBIDE LAMP.

JOHN HAAHTI (CENTER) WITH HIS MINING CREW. NOTE THE MINERS ARE USING CARBIDE LAMPS. ALTHOUGH THE BIGGER MINES LIKE PREMIER CONVERTED TO ELECTRIC LAMPS IN ABOUT 1936, CARBIDE WAS USED ON SMALLER JOBS UP UNTIL THE FIFTIES. CARBIDE LAMPS HAD TWO COMPARTMENTS, THE TOP TANK CONTAINING WATER AND THE BOTTOM WITH GRANULES OF CALCIUM CARBIDE. WHEN THE WATER WAS DRIPPED ONTO THE CARBIDE ACETYLENE, A GAS WAS PRODUCED THAT BURNED WITH GREAT BRIGHTNESS. A FILLED LAMP LASTED MORE THAN TWO HOURS BEFORE HAVING TO BE REPLENISHED, AND WE HAD SMALL TIN CONTAINERS FILLED WITH MORE CARBIDE THAT FIT INTO A HIP POCKET. WHEN THE MINES CONVERTED TO ELECTRIC LIGHTS, MANY MINERS BALKED AT USING THEM, SAYING THE CARBIDES PROVIDE A MUCH BETTER LIGHT.

Chapter Nineteen

JOHN HAAHTI
General Merchant With a Heart

One of Stewart's most outstanding and respected citizens during my childhood was John Haahti, an immigrant from Finland who was best known as the town's general merchant, although he was also a successful miner and prospector.

When his neighbours needed groceries or help, John never could say no even if they could not pay. Many people in the "Dirty Thirties," as well as the '40s and '50s, would have been hungry without John's generosity. Money was not essential — the amount would just get marked down in John's book of credit and you could pay it off when you got back on your feet. John opened the store in 1932 and sold it 29 years later.

John was from a family of nine children, and he ran away from home at the age of 12 so that he could attend free school with the monks in St. Petersburg, Russia. He was fluent in the Russian language all his life.

John's son John Jr. was the delivery boy for the general store, and he was my pal. Once, when John Jr. was out of town, I was hired to make his deliveries. The store's delivery bicycle had a cage attached to the front that was used to carry the box of groceries. One day, I was told to deliver a box to "Louise's house" on Second Avenue. I was about thirteen years old and I had heard about Louise who operated a house of ill repute! Mother had strictly ordered us to stay away from that district where there were evil women who had been known to kidnap children, she had warned! But I rationalized that John paid me a dime for every delivery and I needed the money.

I was very nervous and apprehensive as I approached the house. I grabbed the box, bit my lip, and headed for the door. John had given me the bill for the groceries and instructed me to collect the money. The door was answered by a young lady wearing a kimono, who bade me enter. My heart was beating at double speed.

"Put the box on the sideboard and sit down there," she told me, indicating a wicker chair. "Louise will be down to pay you." She called up the stairs for Louise who presently appeared, tying up her kimono as she came into the room. Louise counted out the money and as she handed it to me, I attempted to rise out of the chair. But the hip pocket of my jeans snagged on some broken wicker, and I felt a rush of extreme panic — Mother was right, I thought, they were trying to trap me! But the wicker gave way, I broke free, grabbed the money and ran out the door to gales of feminine laughter. When I got around the corner, I stopped and counted the money. Louise had given me a 25-cent tip — I was rich! I was never called upon to deliver to Louise's again but was quite sure I would have done it, for money had instantly made me brave beyond belief!

In 1960, John sold the store to Bob Hutchings who renamed it Bob's Mercantile. Four years later, it was sold to Charlie Fenner who still operates it from a newer building on Fifth Avenue. John passed away in 1964 but Mrs. Haahti lived until 1993 to the ripe old age of 92 .

One of the Haahtis' three daughters, Eila, was married and living in Ocean Falls, B.C., in January 1965 when heavy coastal rains caused an avalanche that wiped out part of the village, including the home of Eila, her husband, their two sons and an unborn child. All five died. The people of Ocean Falls commemorated the family with a special stained glass window in the local church. Several years later, as the town of Ocean Falls was struggling to survive, the citizens removed the window and sent it to Stewart's St. Marks Anglican Church on Ninth Avenue where it has since been installed.

ST. MARK'S ANGLICAN CHURCH, BUILT 1910. THE BELL WAS SALVAGED FROM A LOCOMOTIVE THAT WAS LOST INTO THE KITSAULT CANYON AT ALICE ARM.

JOHN'S AD IN 1938, "NORTHERN ARGONAUT."

Chapter Twenty

SAM KIRKPATICK'S HISTORY OF STEWART

1955. FRONT: SAM KIRKPATRICK AND BASIL HOUGHTON. BEHIND: LEGION PRESIDENT JACK CROWHURST AND CHAPLAIN JOHN BOUZEK. BY POLE: JOHN MENEGHELLO

The Kirkpatrick brothers, Sam and Tom, had served as snipers for the Canadian army during World War I and in 1941, during World War II, they formed an Army Cadet Corps in Stewart. The Kirkpatricks were the rifle instructors and John Thompson was the commanding officer. As one of the recruits, I became a pretty good shot. When I reached my 17th birthday, I joined the regular army. During training, we were of course taken out to the rifle range. After a little direction from a young instructor, we were given five shots at a target and I put every one in the bull's eye. The instructor was very impressed and when I told him about my guidance from snipers Sam and Tom, he said: "You should be the instructor here instead of me!" That was the first and last compliment I ever received in my army career!

In his later years, Sam took a job as janitor at the Stewart courthouse. One of his duties was to record the weather, temperature, snowfall and rainfall, but he went further and actually kept a diary of important town events. That diary, from 1945 to 1961, has become an important record of the community's history, as there was no newspaper for those years.

Sam's notes are brief and show no distinction between the levels of importance of these events in the lives of Stewart residents. For instance, the deaths of George Bernard Shaw, the King of Sweden and "Big Chris" of Premier Mines are each accorded a single line, one after the other.

But his personal life surfaces every now and then. For instance, the entry on Aug. 4, 1952 states: "The birthday of Queen Mother Elizabeth and Vi Kirkpatrick." (Vi was Sam's wife.) He later made this mysterious note: "Vi parted with all her teeth." And even later, on Dec. 1, 1953: "16th anniversary of our wedding. Hurray! I dodged rolling pin so far!"

Sam has recorded the marriages of prominent Stewart citizens like Johnny and Dorothy McKenna (proprietor of the drugstore and teacher), as well as births, deaths and disasters like the plane crash on Sept. 11, 1945 that killed resident Margaret Dempsey and her baby. Sam marked the election of Mrs. Sadie Young to the town council on Nov. 30, 1947; she was the first female mayor of Stewart.

His notes indicate a significant number of house fires in that era, and the openings, closures, and re-opening of mines. On most Dominion Days, he remarked on the crowning of this year's Queen of Stewart. Snowfall ranged from nine feet in the winter of 1943-44 to 23 feet in 1946-47.

Some of the other highlights include:

Jan. 24, 1949: Harbor froze over, boat unable to dock.

Aug. 21, 1949: Earthquake tremors (sic) felt in Stewart.

Jan. 4, 1950: J. Broadaxe Coughlan found hanging by his neck.

July 28, 1951: Ernie McMullen died on the trail.

July 19, 1952: Ernie Love drowned with three others in car accident on Hyder Road. Don McLeod, saved.

Nov. 29, 1954: Herb McQuan taken out in strait-jacket.

Mar. 2, 1956: Slim Brown Fire disaster, small children burned, Hyder, Alaska.

Mar. 11, 1956: First robin today, same bird, same date as last year.

July 27, 1956: Four Naval boats in harbor.

June 6, 1958: Hottest day on record, 99 degrees in the shade.

As Sam wound up his tenure, he offered a final, almost verbose, note on his experience as official historian or what he termed "the long List of Jottings at the Furnace Room in the Court House." The punctuation is his.

"A new Era has come, the old one is gone, never to return. The good old days with the old-fashioned crowds; fun-loving, industrious & honest who lived and enjoyed life as it was and asked no more; when locks and keys were unknown, everyone was trusted and were ready with a helping hand; the old time dances were mixed with the modern ones and drew good crowds.

"In summer there were races and sports for adults as well as for children. There were 5 ball teams in the district and the Street Parades were worth seeing. We had a weekly paper* with an Editor who lived and died working for the people and the North Country.

"But Alas, the old class people are gone or going fast. Those that remain and remember are mourners for an Era that was good to the People of the North. THOSE WERE THE GOOD OLD DAYS."

*Sam refers to the Northern Argonaut newspaper and editor Smith. It closed in 1942 and Stewart did not get another until the 1960s.

Chapter Twenty-One

ARROW ROCK

SHORTY CARSON, LEFT, HAVING A LITTLE SNOOZE WITH FRED HASIL (LORD PLUSHBOTTOM) ON RIGHT, 1958.

The old timers of the North always tried to outdo each other with tricks and lies. As Al Lawrence once said: "If you haven't heard a rumor by eight in the morning — - start one!"

The most successful spoof I recall involved Sam McNeil and his sidekick, Shorty Carson. Sam owned and operated a substantially large boat that he rented out for various jobs on the Portland Canal. He liked to bring Shorty along on the trips because he could then be guaranteed lots of laughs!

In the late '20s, Sam was hired to repair some downed telegraph lines on the Canadian side of the canal. The wire stretched for about 40 miles from Maple Bay to Stewart and in order to keep it off the ground, the line was draped over many trees along the way. Sam and Shorty completed their repairs with a long ladder that was always kept on board. One particular day, they finished work early enough to head for the fishing grounds on the Alaskan side at Seal Rocks, about eight miles south of Hyder. The fish weren't biting so Sam had lots of time to think up tricks!

He had an idea and he sent Shorty ashore to pick up a dozen or so weathered sticks, each about two feet long and about the diameter of an arrow. Then they waited for high tide, which, at that time of year, had a 24-foot rise and fall. Sam had spotted a vertical 100-foot bluff just north of Seal Rocks, covered with moss, a few blueberry bushes and contained the odd crevice here and there. When the tide was high, Sam nosed the bow of the boat up against the rock bluff and instructed Shorty to lean the ladder from the boat deck to the cliff. Then Shorty climbed the ladder as high as possible and stuck the weathered sticks into the crevices. With the mission accomplished, they headed for town, rehearsing their story along the way.

In the bar that night, they told their pals that they had met an old Native fisherman that day and heard a story from him about the local Indians. One year, the old fisherman related, when the spawning salmon did not appear, the medicine man told the braves to shoot arrows into the bluff above Seal Rocks to kill the evil spirits that had stolen the salmon. The braves did exactly that and, sure enough, the next year the spawning salmon reappeared in record numbers.

"Well," said Sam, "we ran over to that bluff and sure enough, there were a lot of arrows sticking out." Of course, the boys at the bar didn't believe the story but during the next few months, quite a few saw for themselves the "arrows" above Seal Rocks and it was soon a landmark called Arrow Rock.

SAM MCNEIL

Through the years I have had many a good chuckle when reading Northwest tourist guides that recount Sam's story of Arrow Rock. But David Gunanoot, my good Indian friend, was not fooled. He said: "You think my people crazy, shoot 'em good arrows into the rock? Better use 'em shoot white men!"

Chapter Twenty-Two

GEORGE BUNN: *The Musical Miner*

GEORGE BUNN IS IN THE FOREGROUND WITH A GROUP OF STUDENT HIKERS AT HIS CAMP. NOTE THAN EVEN IN THE BUSH, GEORGE WORE A TIE.

George Bunn was a grand old-time prospector who touched many lives during his time in Stewart.

Arriving in the Portland Canal area about 1910, he brought with him extensive prospecting and mining experience from the Klondike Gold rush of 1898. During his tenure in the Klondike, he had acquired a pretty fair placer claim for which he needed to hire a crew of workers. George was also a great musician who played various instruments, including the guitar. So he refused to hire a new miner to help sink the shaft on his placer claim unless the newcomer was also a musician of sorts! Many evenings, after dinner, they would apparently launch into a genuine old-fashioned jam session that attracted people from all over.

When George moved to Stewart, he staked several hard rock prospects that he worked without success for the remainder of his years. When I was a boy, George lived just a few blocks away and he usually welcomed the company of my friends and I. Because he lived on his own, like most other prospectors, he developed the habit to talking to himself and, in turn, answering his own questions! I remember overhearing some of his conversations and songs that we would hear as we approached his cabin. One of them went:

"George Bunn, George Bunn,
Once a miner, once a mucker,
Now a bum. George Bunn."

Another time I heard him say to himself: "Have a cup of coffee, George." And the answer: "Don't mind if I do, George, cream and sugar please!"

His mining claims were located down the Portland Canal so he needed a boat to access them. One winter, he built himself a rowboat inside his one-room cabin. In the spring, he asked us boys to help him move the boat outside but, try as we might, we couldn't get it through the door. Instead, we helped him tear out part of the wall so we could squeeze the boat through. In payment for our efforts, George played us a few tunes on this guitar!

CABIN AT B.C. SILVER, 1923. B.C. SILVER AMALGAMATED WITH PREMIER GOLD TO BECOME SILBAK PREMIER MINES. GEORGE WORKED AT B.C. SILVER AT ONE TIME.

BILL CRAWFORD AT CASCADE CREEK, 1928.

Chapter Twenty-Three

MIKE THE GREEK

Mike Koris, with his cousin and partner Nick Nickolas, immigrated to Canada from Crete Island about 1920. Almost immediately, they found work at the Premier mine and became miners-in-training. Their dream was to own their own gold mine and toward that end, they staked a claim called Gold River, located six miles up the Georgia River. It empties into the Portland Canal about sixteen miles south of Stewart.

The management at Premier was very good to prospectors and permitted them to take summer months away from their job to work their own claims. The company even provided free assaying for the prospectors. The understanding was that Premier was to get first chance to buy the claims if they amounted to anything, but there were precious few opportunities to test that option!

Mike and Nick bought a small boat for transportation up to the Georgia, plus tools and supplies. Although they were both small in stature, they were strong, tough and determined. They

A VIEW OF MOUNT RAINEY FROM THE HOME OF MIKE THE GREEK.

located several veins and drove several hundred feet of tunnel by hand. They found some gold — never enough to justify a mine but always enough to keep their dreams alive. Just a few more feet, they would declare, and we'll break into the Mother Lode!

Mike would say to me, even in his later years when he was crippled with the backbreaking work of his youth: "Een, just two more pick handles through the dyke and we will be in the good ore!" The old-timers used the handles of the pick, usually about 30 inches long, as a measuring tool in absence of measuring tapes.

In 1929, there was a discovery at the head of the Georgia River, followed naturally by a bit of a boom. Promoters were then trying to tie up adjoining claims, and Mike was invited to travel to Vancouver to discuss the sale of his Gold River claim. But when the "head man" in his Vancouver office showed Mike a prepared sales contract, Mike looked him in the eye and said: "Mister, I no read the English but I tell you one thing" — and he pulled out two 30/30 bullets from his pocket — "if there is any crooked work here, one for you, one for me." The promoter hastily rang for security and asked them to "show this man out, I can't do business with him!" No deal.

Nick died about 1944 and Mike, in his retirement, bought a house in Stewart. Although crippled with arthritis, he still went to his old property every summer and did whatever work he could manage. During the 1950s, I came to know him well and recognized that he had become very homesick for Crete Island. He would say to me: "Een, I no like it here now. Meat no taste, flowers no smell."

A young Greek friend, Tony, moved from Stewart to Kitimat to work at a Chinese food restaurant. His boss, Charlie, talked about a dream to own a gold mine so Tony put him in touch with Mike. Charlie purchased Mike's gold mine for $5,000. Mike sold his home for a few thousand and headed off for his beloved Crete where he passed away four years later.

A couple of years after Charlie bought the mine, I asked him how it was looking. He replied: "Mike the Greek, Mike the Greek, very very smart Greek. Mine thirty years on Gold River, take all the gold to Greece, now nothing left except maybe little bit iron pyrite!"

When I later visited Crete Island, I thought of Mike and fully understood why he had been so very homesick!

Chapter Twenty-Four

ART, THE PROSPECTOR

ARTHUR CAMERON IN FOREGROUND WITH DAVE MENEGHELLO BEHIND AND TEAGUE MCGEE TO THE RIGHT.

Arthur was a professional baker who emigrated from Denmark to Canada to prospect for gold. Born Arthur Fredrickson in Copenhagen, he arrived in Stewart in the early 1930s and when he noticed that most prospectors had Scottish surnames, he promptly changed his name to 'Cameron!'

Arthur was a well-built man of medium height, very muscular and very proud of it! On most winter days, he would break the ice around the deep hole of Rainey Creek and swim amongst the floes just to prove how tough he was. Unfortunately, he paid dearly for this practice in later life — when brother Don and I became involved with him, his kidneys were not healthy and he had to urinate about every 15 minutes. In the bush, he didn't even bother to button up his fly because by the time he had it buttoned, it was time to go again.

Arthur was as stubborn as he was strong. In 1943, he had a contract to drive a tunnel on the Mountain Boy property and as all the fit men were overseas serving in the war, he hired a couple of boys - Don and I - to help him. (We were respectively 14 and 15 years of age.) The wage was $20 a foot, a price so low that we really worked for nothing.

The tunnel was to be driven by hand, working with hammer and steel. An eight-pound hammer was known as a double jack and a four-pound as a single jack. Double-jacking meant two men, one with the hammer and the other turning the steel. Single-jacking was hammering with one hand and turning the steel with the other.

The steel was shaped like a chisel on the end and sharpening and tempering was done with a blacksmith forge and an anvil. The more skilled the miner was in the blacksmith shop, the better and faster was the drilling. The hammer handle was very important, as the miners wanted the hammer to spring back when it contacted the steel. To that end, they spend hours shaving and sanding the handles. (During the July 1 - 4 joint national celebrations for Stewart and Hyder, the hand drilling contests were among the most popular events. In 1936, Roy Becker and Sig Grotto drilled 19 5/8 inches in fifteen minutes of double-jacking. Eric the Bear single-jacked fourteen inches in ten minutes.)

Our work at the tunnel started with the drilling of five three-foot holes — one at the ceiling (the back hole), two chest high (breast holes) and two on the floor (lifters). It was followed by the really ugly work, called "springing." The idea of springing was to make a pocket at the end of the holes, and it took several loads of dynamite to make the pocket large enough. First, we inserted into each of the ends a cap and fuse with half a stick of dynamite, and then we lit the fuses. After the shots went off, we headed back into the tunnel almost immediately, in spite of a thick wall of dust and smoke. We used a special tool to scrape out the small pieces of rock from the hole ends, and then we blasted again, using a whole stick and a half of dynamite. The third time we would use three sticks and usually after that blast, the pockets would be able to hold a case of dynamite. We then went home after lighting the fuses for the fourth time. (In those days, a case of dynamite cost about $5.00 compared to its current value of about $150.00.)

A later adventure with Arthur, when he and I drove the Gold Drop tunnel, resulted in an accident. At the end of one day, after we had loaded the last blast, he sent me out to do some chores at the portal. It was evening and getting quite dark. I stood waiting for Arthur after the chores were done when the shots began to go off and

ART'S EXPLORATION CAMP ABOVE BEAR VALLEY

the smoke started to billow out of the portal. Worried about what had happened to him, I was just starting to head inside and search, when he showed up in my headlamp. He emerged, crawling on his hands and knees, and when I pulled him over to sit on a log, I saw that his face was all bloody and dirty, and his hat and lamp were gone. I managed to lead him to our camp, which was a mile away. While I was bandaging him, he told me what had happened. Because we were drilling in wet ground, he had wanted to waterproof the joint between the silver cap and the black fuse, so he dipped it in tar. He had just finished lighting the fuses with the "spitter" (an eight to ten inch length of fuse that is split every half inch in order to light the main fuse) and was about to throw it away and head out when an unnoticed tarred cap attached to the spitter exploded in his face. The explosion knocked him off his feet, blowing away his hat and carbide light, but he managed to crawl around the corner of the tunnel before the round went off.

Arthur was very ashamed and swore me to secrecy. He wouldn't even go to see the doctor in town and as a result, lost the sight of one eye. It was a lesson for me; I decided to try different lines of work and especially a different boss!

Brother Don tells of working with Art on the Gold Boulder claim, where the two of them bunked together in an old tent. It was October, about the time of the coastal monsoons, and the old tent leaked badly. Don and Art build a canopy over their beds with saplings and covered the canopy with some tarpaper in order to keep their beds dry. Art, naturally, kept a tin can under his bed into which he relieved himself every fifteen minutes or so. One night soon after they'd gone to bed, Don heard Art laughing. "Ho, ho, I've got a canopy over my bed and a can of pee under my bed!"

Years later, when we were working in the big mines, Don put our experiences with Arthur in the right perspective. "What we learned with Arthur has stood us in good stead because when we came up against a problem, all we had to do was figure how Arthur would do it and then we'd do the exact opposite. It worked every time!"

Art also had a knack of getting lost on his prospecting trips, and I was twice called on to search for him. On both occasions he was carrying the pack that contained the food, while his partner carried the tent and sleeping bags. And both times Arthur fell off a log while trying to cross a creek, naturally letting go of the food pack and making the remainder of their trek a rather hungry one.

In the second incident, his partner Bowser Bill made it into Stewart and told Julian Berkosia and myself where he had left Arthur. Julian and I travelled a hard day before we found him, too crippled to walk and on the far side of a creek that was too wild to wade. We could see his fire, and yelled and threw rocks until we got his attention and he crawled out of the bush. We threw some chocolate bars across, which kept him while we could hike miles upstream to a suitable crossing and then back downstream to his camp. But Art was in no condition to walk out, even with our assistance, so I headed back to town for horses. Julian stayed with him and was mighty angry the next morning to discover that, overnight, Art had eaten the entire pack of grub we had brought and smoked all the cigarettes! I arrived with horses the next day, and we were all back home by nightfall.

Arthur died a comparatively young man.

Chapter Twenty-Five

MARMOT RIVER CASEY

AUTHOR IAN AND CASEY, 1960.

He claimed his name was Albert Raymond Casey. He said he was born in Fredericton, New Brunswick; his father's name was Aloysius and his mother, Mary. When his friends in Stewart decided Casey must be over 70 years of age and therefore eligible to receive the Old Age Pension, I spent a lot of time trying to establish who he was and where he came from. The pension authorities needed documented proof as to his age, but I was stymied at every turn.

Casey would just chuckle and say, "McLeod, you're not very good at backtracking."

Being a trapper and prospector, he sure was

good at covering his tracks. Vital Statistics had no record of him so I next tried the parish priest in Fredericton because Casey said he had been born a Catholic. I received a delightful reply from the priest who said he was 91 years old himself and had no record of Casey and furthermore did not recall any Casey family in Fredericton.

Casey used to talk of Greece and Turkey and the Dardanelles campaign of the World War I. He confirmed that he had been a soldier, and rattled off some numbers when I asked if he remembered his regiment. So I sent his name and number to the Department of National Defence in Ottawa, who replied they had no record of him.

By then, I was becoming suspicious. When I asked where he went to school, he listed places like Kansas City and Fairbanks, Alaska. I waited a couple of weeks and asked again for his regimental number and all the places he went to school. New number, new towns!

At my wit's end, I happened to mention the quandary to Al Lawrence, the secretary of the Moose Lodge who said, "You know, I was going through the old records in the office awhile ago and I'm sure I saw Casey's name. I'll see if I can find it." Sure enough, Al found an application by Casey to the Lodge in 1925 that gave his age at 35. I sent the document to the government and he was finally granted a pension. Casey concluded that I was a better backtracker than he had thought!

Casey was well known around town, and the locals were quite fond of him. But his background was clearly a mystery and he didn't ever seem anxious to clarify the situation. Naturally, after my vital statistics adventure, I was a little more curious than I'd been before. Casey was a regular in my bar but generally handled the liquor well. One night I was bartending when he had a larger than normal snootful. The next day he arrived at the bar looking pretty bedraggled and called for a cool beer. He confessed he was feeling "terrible." So I sat down beside him, thinking this might be a good time to trip him up. I said: "That was quite a story you told me last night about killing that guy you caught with your wife." To which Casey replied: "Boy I sure must have been

CASEY AT SWAMP POINT CABIN, 1958

HALF PRICE BOOKS®

Half Price Books
1835 Forms Drive
Carrollton, TX 75006
OFS OrderID 39528958

Thank you for your order, Main Address!

Thank you for shopping with Half Price Books! Please contact service78@hpb.com. if you have any questions, comments or concerns about your order (113-9928434-8051426)

Visit our stores to sell your books, music, movies games for cash.

SKU	ISBN/UPC	Title & Author/Artist	Shelf ID	Qty	OrderSKU
S438296635	9780973516500	Prospectors Promoters and Hard Rock Miners McLeod, Ian; McNeil, Holly	AH4.1	1	

SHIPPED STANDARD TO:
Main Address
800 Avondale Ave
1033841-1-0568
Grandview Heights OH 43212
zypxchqnwm9p8bz@marketplace.amazon.com

ORDER# **113-9928434-8051426**
AmazonMarketplaceUS

drunk last night." End of my attempt to pin him down!

When I first met Casey, he had a cabin on the Marmot River because his claims were upstream. When they didn't pan out, he moved to Coon Bay, about 20 miles further down the canal in order to be closer to his trapping and fish poaching. He used an open boat with an outboard motor in both winter and summer. Even when he was in his 70s, he would come up the canal in the open boat during the winter, spray freezing all over him so that by the time he got to the hotel his mustache would still be frozen. Hot rye was his drink, first thawing the mustache and then his throat.

Although he lived alone in the bush all his life, he was gregarious and a great talker. If he didn't have any body to talk with, he chatted to himself. I overheard him many times and the conversations were always intelligent. Casey had a battery radio in his cabin and so he was always up to date on world affairs.

Casey was a man of medium height, slim and he had extraordinarily white front teeth in spite of the fact they had never met a toothbrush! He considered himself a real ladies' man right to his dying day, and it was true that the ladies appreciated his charm and gentlemanly manners.

The Portland Canal is 60 miles long and averages about one and a half miles wide. The west side is Alaska territory and the east is Canadian with the boundary line up the centre. Casey's trap line traversed the upper 40 miles on the Canadian side but as no one ever trapped the American side, Casey would run traps there as well. He would say, "I know Uncle Sam doesn't mind me harvesting his fur because otherwise the area would get overpopulated."

The Portland Canal was closed to commercial fishing in the thirties and forties, but in 1952 the Alaskans opened their side to gillnetting although the Canadian side was still closed. The stream beside Casey's cabin had a good salmon run and when he saw all the dollars swimming upstream, his Irish poaching genes kicked in (so he said)! He had a gill net and made a deal with an American fisherman to buy his fish. Soon he was really in the money. The fish were Chum salmon that were commonly called 'dog salmon.' The Americans would say, "Casey is selling us the Queen's dogs."

Canadian Fisheries officials knew Casey was up to something but he was too canny to be caught. So the next spring they offered him the job of Fisheries guardian complete with a salary and boat. The old poacher was now looking for the other poachers but he never did arrest anyone!

Frank Calder, a native from the Nass River district, was our long-time Member of the Legislative Assembly. Casey had an election sign nailed to his door with a picture of Frank, who was a member of the CCF party, and the words, "Vote for Calder." One of the boys questioned Casey about the poster. "I thought you said you were a Liberal but you have a CCF sign on your door." Casey replied: "When I'm in town I'm on the Liberal team but down the canal I'm on

Calder's team. When the Nass Indians go by and they see Frank on my door they never bother me."

I once overheard Casey discussing with a tourist the subject of trapping. The visitor asked if Casey ever tamed any of the wild animals, to which he replied: "Yes, I have a pet wolverine."

"But aren't wolverines very vicious? Surely you can't pet it?"

"Yes, I do."

"How in the world can you do that?"

"With a long, long stick."

Casey started telling the boys that he had a live-in girlfriend down at his cabin. Sonja was her name and she was from Bella Coola, apparently, but she didn't like visitors so Casey told them never to come around. Some of the fellows believed him but most of us did not. One of the non-believers went fishing one day and he took along some of his wife's old clothes, stopped at Casey's cabin and hung the clothes on the line. Afterwards, there were more believers than not, and John Lehto, Casey's old friend, was absolutely certain that Sonja existed. But when the same clothes, considerably weather-beaten, were still on the clothesline next spring, the believers admitted that Casey had fooled them again.

Casey was asked how he came to be so healthy at the age of 85. He said, "It must have been all the bear fat and the beaver meat I ate but that rancid bacon almost ruined me, so I quit it."

To prospectors, a 'deal' refers to an agreement to sell their claims to a promoter or mining company. Casey always talked about deals. A good deal could be anything from a girl smiling at him to one of the boys buying him a drink. A bad deal could be anything from bad weather to bad whiskey. And if things went well, he would always say, "That's a good deal!"

Casey passed away in 1984 at the age of 96 and was active up until a few short days before his death. Those of us who knew him remember him with great fondness.

JOHN LEHTO, MANOT, CASEY

Chapter Twenty-Six

BLACK NICK

Nick Giloff was known by all as Black Nick. He worked as a miner at Premier for many years and retired to Stewart in the early 1950s. Black hair, large black eyebrows and a dark complexion were the features that led to his nickname. He was very husky and strong in his youth, but was a pleasant and gentle man. Black Nick was born and raised in Russia in the same district from which Stalin hailed, and Nick was very proud of the Russian leader.

Later in life, Nick sold his shack in Stewart and moved to Vancouver, where he rented a room on Cordova Street. I would visit Nick whenever I went to Vancouver. One of our conversations went thus:

"How do you like Vancouver, Nick?"

"I like it pretty good, no snow to shovel and I like to walk around on the nice sidewalks. Sometimes, I walk all the way to Stalin Park."

"Nick, that's Stanley Park, not Stalin Park."

"Yes, I know, Stalin Park."

Lord Stanley would not be amused.

PHOTO BY LATE DR. BILL KERGIN

FORM 18 5M 11-24 SELBY PRINT

LOT 6257 San Francisco,

RETURNS OF Ore RECEIVED OF Porter Idaho Mining Co. BY
Stewart, B. C.

Copy

AMERICAN SMELTING & REFINING CO.
SELBY SMELTING WORKS
444 CALIFORNIA STREET

Mark S. S. "Doylestown"

Arrived 8/28/25

1432 Sacks weighing gross	115,660		lbs.
Samples			lbs.
Tare	1,900	113760	lbs.
Moisture 6.2 %		7053	lbs.
NET WEIGHT		106707	lbs.

ASSAY PER TON		
Gold	.06	ozs.
Silver	123.00	ozs.
Lead (wet)	8.3 -1½- 6.8	%
Copper (wet)		%
Insol.	66.6	%
Iron	7.2	%
Zinc	1.1	%
Sulphur	1.6	%
Antimony	0.2	%
Arsenic	0.7	

PERCENTAGES AND PRICES	DEBITS		CREDITS	
Less 5 %@ 20.00			1	14
Less 5 %@ .72 3/8			84	57
Less 10 %@ 7.935 per lb. Less 1½ 6.435			7	88
Less % per lb. Less				
Treatment	12	00		
Totals	12	00	93	59
Value per Ton			81	59
Value of 106707 lbs.@ 81.59 per Ton			4353	11
Sampling and Assaying				
Reducing				
Burning, Grinding and Sifting				
Freight at 6.00	346	98		
Insurance Tacoma to Selby	25	00		
R. E. Chase & Co.	20	00		
Custom House Entry	4	50		
Totals	396	48		
NET PROCEEDS			3956	63

SAMPLES	GOLD	SILVER
Orig.	.06	123.50
Dup.	.06	122.50

Settlement made on

Quotations of

THIS IS THE SMELTER RETURN FOR 57 TONS OF ORE SHIPPED TO THE SMELTER IN 1925. THE SACKS OF ORE WOULD HAVE BEEN HAULED BY PACKHORSE 9 MILES BY TRAIL TO TIDEWATER AND THEN BY STEAMSHIP FROM STEWART TO THE SMELTER. WITH THE COST OF MINING, SORTING, SACKING AND TRANSPORTATION APPLIED TO THE PROCEEDS THERE WOULD NOT BE MUCH PROFIT LEFT. PRESENT DAY METAL PRICES WOULD GIVE TEN TIMES THE 1925 VALUE.

Chapter Twenty-Seven

JOHN LEHTO

JOHN LEHTO ON FIFTH AVENUE IN 1998.

When John Lehto passed away in 1998, at the age of 90, I thought to myself, "There goes the last of the old time prospectors."

John arrived in Stewart from Finland in 1927, the same year I was born, so he always said, "Eon, you and I came to Stewart the same year so we are both old-timers."

He was strong and tough and spent his entire life prospecting and trapping. John was one of the few men I knew who traveled regularly by foot or snowshoe over the rough glacier country between Stewart and the Unuk River valley.

John had copper claims in the Unuk and of course they were for sale, as most claims usually are. Eric Ostensoe, the field geologist for Granduc Mines, examined John's property once and decided the claims were interesting enough to warrant some work. On behalf of his company, Eric made an offer to purchase. John took the written offer home to study. It was a standard contract for the times, something like $5,000 down, $5,000 a year later and by year five, $100,000 plus a small royalty interest.

The next morning, Eric and John met in my coffee shop where I was on hand, at John's request, to explain legal language to him.

Eric asked, "Well, how does the deal look, John?"

John replied, "Very nice deal, Eric, very good, only one thing wrong."

Eric: "What's wrong with it?"

John: "Payments are upside down." No deal.

But Eric wasn't the only potential buyer to come John's way. He had several properties and from time to time, mining geologists would express interest. One day I saw John with two strangers and when we met later, I asked him who they were.

He replied, "Just a couple of ham and eggers."

"What do you mean by that — ham and eggers?"

"Geologists who came up to my cabin the other night to look at my claims. I made them coffee and hardtack for breakfast but they said no, they must have a good feed of ham and eggs before we tackle the mountain. So that is why I call them ham and eggers."

I told that story around and soon many of the boys were calling geologists "ham and eggers." Even my son Rob, a geologist himself, started calling himself and his associates by that term. So if you ever hear someone talking about ham and eggers, you'll know where the story began!

John was a loner who never had a partner, but the work he accomplished was unbelievable. He drove hundreds of feet of tunnel by hand — that entails drilling holes with hammer and steel, blasting, shoveling the broken rock into a wheelbarrow, pushing the barrow to the surface, and dumping it.

Like so many others, John believed that the next round he blasted would expose a wider and richer ore body, and like most of those others, it never happened. When I think of John and the others, I think of the old Cousin Jack's saying about ore: "Where she be, there she be," and wonder if his saying shouldn't be changed to: "Where she don't be, there she don't be." But prospectors are a breed of optimists who are even more determined than promoters. Once they locate a vein, they are certain that it's just the start of a big rich ore body. As Jack McBeath would say: "They think it's just showing its teeth." Not often would their optimism be rewarded but when it was, the benefit was not just to the prospector but also to the community and the nation. Eskay Creek is a good example of what a persistent prospector can accomplish.

John Lehto was a lifelong bachelor but he did like the ladies. In later life, he twice became engaged but in both cases, the ladies died before the planned weddings.

Chapter Twenty-Eight

PARTNERS

AUTHOR IAN MCLEOD WITH PARTNER ANGELO BUGNELLA

When I went into business in Stewart in 1952, I just couldn't get mining out of my system so I started spare-time prospecting. My main partners were Angelo Bugnella and Ramo Jokanovich.

Angelo, an Italian from the Tyrol, was a slight man who could walk like a mountain goat on the hillsides. Ramo was from Yugoslavia and was husky, strong and a good worker. At times, we had Arthur Hagfor, a Finn known as Happy the Trapper, who as a topnotch bushman and David Gunnanoot about whom I have written in another chapter.

We concentrated our efforts on upper American Creek where we had a few interesting showings. Diamond drilling, however, proved the ore had no depth. We had two cabins, one at a place called Bear Camp and the other well above the timberline built from lumber that was flown in by helicopter.

Bear Camp was well named, as there were lots of animals around, including both blacks and grizzlies. Although most prospectors hated bears due to the damage to cabins and caches, we decided to try to live in harmony with them. Our little Bear Camp log cabin had one window and a door, and the bears would break in through both openings. I had been told by an old-timer to put chicken wire on the window and hang a crosscut saw on a rope in front of the door. That seemed to work for a few months until one day we arrived at the cabin to see the saw still hanging but the door smashed. The window was gone, and the chicken wire was broken and bent outwards, indicating that the bear must have entered by the door and exited by the window!

PARTNER RAMO JOKANOVICH

One morning at Bear Camp as we were preparing to hit the trail, we heard Angelo swearing on the porch. We asked what was wrong, to which he replied (expletives removed!): "I left my boots on the porch last night and the porcupine ate them up. Just the soles left. I can't go up the

A PROSPECTING CREW AT EALU LAKE. FROM LEFT, BOB WATSON, JOHN ANGUS MACDONALD, ROCKY LAROQUE AND AUTHOR IAN MCLEOD.

hill today." Porcupines have strange appetites. They love plywood, hydraulic hoses and anything salty. They likely thought Angelo's smelly footwear was quite a treat!

Bears seem to recognize that cans contain good food. The great beasts are quite capable of opening cans of food and removing the goodies without cutting themselves. In our upper cabin, above the timberline, it was quite cold and so we had to outfit it with an oil space heater that was about a foot square and three feet high. Once we returned to discover the oil stove outside, smashed to pieces. Undoubtedly, Mr. Bruin must have thought he had hit the jackpot — the promise of a huge chunk of ham in a can that size!

Both Ramo and Happy the Trapper were dead shots with their rifles. One day, the two were walking together toward Ramo's cabin near Glacier Creek. As they rounded the bend below the cabin, they came face-to-face with a grizzly that was just leaving by a broken cabin door. Both prospectors raised their rifles quickly and shot at the same instant. The bear fell dead. Upon investigation of the carcass, they saw that one bullet had hit between the eyes and the other into an eye. Ramo pointed to the one in the dead center and declared: "That one's mine!" Happy chuckled when he told me about it later — he was just happy that one of the bullets had done the job.

PARTNER HAPPY THE TRAPPER

Chapter Twenty-Nine

TONY MARIOTI

TONY MARIOTI, 1958

Guiseppi Marioti was born in Monte Cattini, Italy, in 1890. He went to school for four years, worked in the fields until he was 18 and saved enough money to buy a one-way ticket to America. Here, he was known far and wide as just "Tony", one of the more interesting and colorful characters to live in Stewart.

Tony and our father were best friends and after Dad passed away, brother Don and I came to know Tony very well. He never quite conquered the English language and was inclined to put an "o" behind most words. He called me "Eeno." He would say, "Eeno, before he die Jacko McLeodo and Tony were besto friends and now Eeno and Dono McLeodo my besto friends." Tony was very gregarious, loved to entertain with the odd song and dance, and was a great storyteller. I remember two favorite Tony stories that I pass along to you.

In 1914, Tony was out of a job, broke, and down to one sandwich in his backpack. He rode the rods to Revelstoke, B.C. and was wandering the streets when he happened to pass by a schoolyard. In the yard was a group of men with broomsticks on their shoulders marching back and forth on orders from an officer. Tony made his way toward the officer, to whom he said: "Misto, I'm looking for a jobo." The officer handed him a broomstick and told him to get in line. So Tony got a very good jobo — working in the trenches of France for four long years!

Tony was a miner so the army put him in the engineer corps and when he got to France, he served in the tunneling company at various locations, including Vimy Ridge. He often spoke of his good friend, Sgt. Browno from Nova Scotia. He was Tony's kind of man — strong, brave and fair. One morning, when the troops were in a rest area and lined up for parade, the Colonel asked for two volunteers to sneak behind the German lines to blow up a bridge. Sgt. Browno stepped forward and so did Tony. Tony told us: "I would go to hell for Sgt. Browno." So that night the two of them gathered a load of explosives and set off across no-man's land. They eventually found the bridge, successfully blew it to smithereens and returned safely to camp. Soon after, they were lined up again to receive the Military Medal for Bravery. As they waited, side by side, for the officers to reach their spot in the line, Tony noticed a slight, small officer leading the group and whispered to Sgt. Browno: "Who is that little punko coming here?" The little punko was the Prince of Wales, later King Edward VIII, who pinned the medals on the men and shook their hands.

After the war, Tony was discharged in Vancouver and, with his mustering-out pay in his pocket, he decided to start a restaurant that would specialize in Polenta, an Italian dish made with cornmeal. Business was looking very good in the beginning: men lined up to eat at his joint. But then word got out that you didn't have to pay Tony for your food until payday and for most of them, payday never arrived. Tony had to abandon his restaurateur dreams and head back for the mines.

The Alice Arm area was booming in 1919 with the Dolly Varden mine starting up. Tony met our dad there and they became the best of friends. The following year, they moved to Stewart and that's where both of them spent the rest of their lives.

By the late 1940s, Tony had to quit mining as he was loaded with silicosis and also suffered from a severe skin ailment. By that time, he owned a shack in Stewart and we young fellows enjoyed visiting his place for homemade wine (from real California grapes) and homemade chicken cacciatore.

But Tony's skin problem was painful. One day he said to me: "Eeno, I go many times to the veteran's hospitalo but they can't help me. I go to Dr. Greeno in Prince Ruperto and he is the besto doctoro in Canada but I'm not getting better. I talk to an Italian guy the other day and he says the besto doctoros in the worldo are in Roma. I go to Roma."

Money was not a problem. Tony had a good silicosis pension as well as an army pension so I arranged for a return ticket to Rome and away he went.

Some days into his trip, I received a telegraph

from Tony in Rome. (There was no long distance telephone service in Stewart at the time.) "Urgent. Send Copenhagen snuff right away." Tony was addicted to snuff so I packed up 24 cans and shipped them off by airmail. When Tony arrived home, he advised us that Rome was a very backward city as there was no place that sold snuff and, as for the Italian doctors, they did not know as much as that Scotcho, Dr. Greeno in Prince Ruperto!

One day, an excited Tony told me that a salesman had sold him a subscription to an Italian magazine. He could not read or write English but could get by in Italian. He said: "Eeno, it's too bad you can't read Italian because the magazine is about Christian democracy and it's got lots of common senso in it."

"What is the name of the magazine?" I asked.

"It's called Wake Up."

I realized it was the Awake magazine published by the Jehovah's Witnesses church but I didn't say anything. Months later he came to see me in a very agitated state. "Eeno, that Wake Up magazine is no good. I just got one and I burned it. Write them and tell them not to send it any more."

"What was wrong, Tony?" I asked.

"They said Mussolini was no good but he was the besto man because he put all the people to worko in the Depression." That was very important to Tony who had experienced being out of work so often in his life.

During the late 1940s, Bob Hutchings had bought the dairy from John Campbell, including a bull and about a dozen cows. When one of the cows gave birth to a male calf, Hutch donated it to Tony who planned to pasture it during the summer and butcher it in the winter. Tony named the calf "Bullo" and was so kind to him that Bullo would follow Tony around town like a dog, even waiting outside the bar when Tony would stop for a drink. When fall arrived, Tony couldn't possibly think of eating his friend so he decided to raffle him off. Within a week the tickets were all sold and everyone headed for the bar, where the winning ticket would be drawn. The winner was a lady from Prince Rupert. She was a good sport and rode Bullo through the bar in celebration. She then telegraphed her husband in Prince Rupert: "Have won a bull. What to do?" His reply was: "Peddle your bull and come home." She did, and the new owner was hungrier and less softhearted than the others, so poor Bullo became hamburger.

Provincial regulations of those days required fishermen to have fresh water licenses (not necessarily salt water licenses), but the laws were seldom enforced. The only trout in the area were Dolly Vardens that fed off the eggs of the spawning salmon, and it was a common belief that catching Dollys would benefit the salmon run. One day, Tony and Tommy Palmer decided to go fishing to Clements Lake and they soon had a nice batch of trout. As if on cue, the new game warden appeared and asked them both for their fresh water licenses. Of course, they had none and so he arrested them. Tony thought it patently unfair to start enforcing laws that had for so long been disregarded, and he vowed he would go to jail rather than ever pay a fine.

Court day arrived and the two culprits appeared before the lay magistrate, Billy Orr, who was notorious for never finding anyone "not guilty." So, in spite of their arguments, the accused were found guilty and fined $25 or one week in jail. Tommy paid his fine but Tony was adamant that he was willing to be jailed. The local policeman tried hard to convince him to pay the fine because, as well as being sympathetic, it was also true that the policeman's wife would have to provide meals for a prisoner. But Tony stuck to his conviction and into the cell he went. The policeman didn't lock the cell and soon his family and Tony were the best of friends, with Tony cooking spaghetti dinners for them and even babysitting. While in custody, Tony was permitted to go downtown to buy his Copenhagen snuff but not allowed to go to the bar — which he said was fine, because he was saving money!

By 1959, Tony's silicosis was very debilitating so we shipped him off to the Veteran's Hospital in Vancouver but he died two weeks later. He was a strong brave character and I still miss him.

TONY'S FRIEND OWNEY (BULLSHIT) McFADDEN, 1928.
OWNEY LOVED DOGS AND HORSES.

Chapter Thirty

ALPHONSE AT TIDE LAKE

CHARLIE WALKER IN LEAD AND ALPHONSE IN REAR LEAVING BIG MISSOURI WITH SUPPLIES FOR TIDE LAKE

Alphonse, Premier miner and prospector, was described by locals of the day as a "real" Frenchman, indicating that he came from France not from Quebec.

Alphonse staked his claims in the Tide Lake area on the west side. Tide Lake was originally six miles long, a mile wide and 500 feet deep. Dammed over time by the Frank Mackie Glacier, a natural diversion had been directing some of the lake water into the Bowser River for many years until 1931 when it broke out permanently. Because the water level varied a lot, it was named Tide Lake. The phenomenon that causes these self-dumping glacier lakes is called "jokulkaup."

During the permanent "breakout" of Tide Lake in '31, the released water scoured the Bowser valley, flowed into Bowser Lake, and headed 100 miles down the Nass River to Observatory Inlet. Commercial fishing season was in full swing and the fishing nets sank from the weight of extra silt. The fishermen blamed the Anyox smelter but analysis of the silt showed it was not Anyox. More likely, the source of the silt was the billions of gallons of water released from Tide Lake.

"Jokulkaup" caused much trouble for prospectors like Alphonse who had to get from the trail on the east side of the lake to the west side. To find enough timber to make a raft, one had to climb uphill past the old high water line and then drag the logs down to the water.

After 1931, Alphonse was able to build a pack horse trail to his property, and every year he drove the tunnel on the vein a little farther, always confident that with a few more pickhandles he would hit the bonanza ore!

In 1948, my brother Don accompanied a geologist to examine Alphonse's property. When they arrived at the camp, Alphonse was so happy to see them he broke out a bottle of rum. When the bottle was empty it was bedtime. Alphonse was up again at dawn, making coffee and mixing up a batch of flapjacks, and the boys tied into it, as they hadn't had dinner the night before. As they poured the syrup on the pancakes, they noticed little black specks that turned out, on closer inspection, to be hundreds of tiny ants! When a subtle glance at Alphonse revealed that he was chewing happily, ants and all, the other two quietly scraped the syrup off the pancakes and finished their breakfast.

Alphonse's job at Premier was as dump man, working at the ore bins on the dock. In the mid-1940s, the company had shut down the aerial tramline and started hauling concentrate by trucks. One of the drivers was my friend Bill Michaels and Bill told me a story about Alphonse.

One day, Alphonse told Bill: "I got a stew on the stove that's been simmering for a week and it's got the best flavor you've ever tasted." So he took Bill to lunch in his shack, which was located a short distance from the dock and close to the Alaskan border. It was a typical prospector's shack with a bed, table and chairs, and a stove. Above the stove was the inevitable short wire line that would be used to hang clothes to dry.

When they got to the shack, Alphonse put more wood on the fire and before long the large pot of stew was bubbling. Alphonse ladled out two plates of stew that not only smelled wonderful but also tasted great. Hungry Bill cleaned up his plate quickly, and immediately was offered a second helping. But while it was being served up, a large object appeared on the ladle. Said Alphonse: "My goodness, I've been looking for that sock for a week!"

Alphonse died on the trail to his cabin soon after the sock incident but Bill lived for many years after so it wasn't likely the sock that killed the old prospector! At that time, the local undertaker had recently passed away himself so my friend Bob Hutchings volunteered to do the job although he was definitely a novice in this area. He took the body to the morgue and asked for help from Dr. Burtnick. Together with a bottle of scotch, two bottles of formaldehyde, and a textbook, they proceeded to embalm Alphonse. Hutch said that by the time they finished, Alphonse's cheeks were as rosy as a young boy's, and the two embalmers needed help from each other just to stand. By the way, that was Hutch's first and last attempt at undertaking.

Alphonse loved to tell us: "What this country needs is a good crooked promoter to get a mine going." I'm not so sure he wasn't absolutely right.

Chapter Thirty-One

THE KRISTOVICH FAMILY

Nick Kristovich was from Yugoslavia and he married a native lady from Port Simpson, B.C. Nick staked a home site at Tombstone Bay on the Alaska side of the Portland Canal, about 40 miles south of Hyder. The couple had five sons and a daughter.

Nick had been a boat builder in the old country and he built a 56-foot seiner (fishing boat with special nets) at Tombstone with the help of his sons. The ship was called "Five Brothers" and it was a fine vessel from which son John fished for many years.

Nick also kept a house in Hyder so that his sons could attend Stewart school, and that is where I came to know his sons Pat and John.

Pat swelled the Kristovich fleet when he bought the "Blue Eagle" seiner. In 1953, I went with Pat on the Blue Eagle to the Nakat Inlet, 60 miles south of Hyder. We anchored and left two friends on board while the two of us headed inland for a little hunting. While we were away a strong wind blew up, dragging the ship's anchor onto the rocks. The two men on board didn't know how to start the engine or what to do; but as the ship was sinking, Pat managed to get on board and send out a distress signal that was heard by his brother John on the "Five Brothers". In a few hours we were all safely aboard that ship.

The Blue Eagle was salvaged and repaired, and stayed in business for many years.

IN 1953, DURING ONE OF PAT KRISTOVICH'S FISHING TRIPS, HE CAME ACROSS DOUG HAIG AND DON NORTON SITTING ON THE UPSIDE-DOWN PONTOONS OF THEIR AIRCRAFT ABOUT 20 MILES SOUTH OF HYDER. THE PLANE HAD FLIPPED DURING LANDING, SO THE BOYS WERE SURE HAPPY TO SEE PAT, WHO LATER DECLARED: "THIS WAS THE BIGGEST FISH I EVER CAUGHT!"

Chapter Thirty-Two

THE STRANRIER FLYING BOAT, "SKEENA QUEEN," IN STEWART HARBOUR, 1945. OPERATED BY THE QUEEN CHARLOTTE AIRLINES, IT WAS HAULING FREIGHT TO MORRIS SUMMIT MINE (LATER CALLED SCOTTIE GOLD.) THIS AIRCRAFT WAS LOST WITH ALL ABOARD ON A MERCY FLIGHT LATER THIS SAME YEAR.

PILOTS AND THEIR CREWS

Although there were itinerant aircraft in the area from time to time during the 1920s and '30s, it wasn't until 1945 when Jim Spillsbury's Queen Charlotte Airlines (QCA) opened a base at Stewart. They put into service war surplus Stranraer flying boats to help with the early exploration at the Morris Summit mine and to haul freight to Summit Lake. About the same time QCA had also started a twice-weekly schedule flight from Prince Rupert to Stewart using Norseman aircraft. Their first agents were Jack and Muriel MacKay, and Muriel carried on alone for many years after Jack's death. Muriel transferred to Pacific Western Airlines when they bought out QCA and stayed on the job until her retirement in 1975. Muriel was very well known and well loved by all the pilots, many of whom considered her like a second mother. Her distinctive call sign was "Rupert, Rupert, Stewart" and it is also well remembered.

HSL FLYING BOAT AT STEWART IN 1922. THE PILOT WAS EARL MACLEOD WHO BECAME AN AIR COMMODORE IN WORLD WAR II.

When Granduc mines started exploration in the early '50s, they bought their own aircraft. Both the Beaver and the Super-cub were equipped with wheel skis that enabled the pilots to land by skis on the glaciers year round and at the Stewart air strip on wheels. The Beaver, piloted by Harold Kellough, crashed and burned at the toe of the Leduc Glacier in a snowstorm. Harold was the first Granduc fatality. The Beaver was replaced and the job went on.

Emerson (Wally) Wallace was hired to run the show. Now Wally was a bit overweight, especially around the middle. While flying the Super-cub over the glacier one time, he got caught in a snow squall and landed on the snow-covered glacier. As he cut the power back the skis caught in a crevasse and the plane nosed down into soft snow. Wally was thrown through the windscreen, straight toward the propeller that was still turning strongly. Fortunately, Wally's big rear end wedged in the broken glass and stopped him only a few inches in front of the propeller's blades, literally saving his life. After he left Granduc, Wally flew for the rest of his career around Vancouver Island.

In the late '50s, Granduc mines undertook winter Cat train freighting from Nine Mile over Salmon Glacier and on to their camp on Leduc Glacier. The two glaciers were connected. By late spring, the Cat trail became impassable so aircraft were needed to continue the freighting. They first tried using float-equipped aircraft loading from a barge at Burroughs Bay in Alaska, taking off from the water and using the pontoons to land on the snow on the glacier. Two Norsemen and a Seabee were used but the experiment was not successful. As the summer season progressed and the glacier receded with the warm weather, the planes had to land further and further up the glacier and away from the camp. The solution was finally to build a gravel airstrip on the side hill near camp and use wheel-ski Otter aircraft from the Stewart airstrip. This method of service was used until the two ends of the mine tunnel were connected in 1968.

Bill Ross, originally of Scotland, began his career as an air engineer in the Royal Air Force during World War II. When the war was over, he immigrated to Canada and continued working at his trade. I call him the old bold Scot! He came to Stewart in 1955 as base manager and engineer for Trans Provincial Airlines. He later moved on to Pacific Western in the same capacity and stayed with them until that air service ceased in 1975. (The opening of Highway 37A, connecting Stew-

SEABEE AIRCRAFT OPERATING ON THE GLACIER BY GRANDUC

➤
NORSEMAN AIRCRAFT OPERATING ON FLOATS ON THE GLACIER AT GRANDUC

art to Terrace and Smithers by road for the first time, caused the demise of passenger air travel out of Stewart.)

Bill was front and center when the large tonnages of freight were flown from Stewart to Portal Camp on Leduc Glacier. Many types of aircraft were used but mainly single Otters and C-46s. The latter could haul up to seven tons and land on their wheels onto a glacier air strip that had been compacted by D-8 caterpillars.

That was about the time Bill was involved in the first "sky jacking" of northwestern BC! The weather was socked in one day so Jack Magel, the captain of one of the C-46s, went fishing in the harbour. His plane was loaded, though, and the supplies were needed at the mine. When the fog blew away, revealing a perfect flying day, Jack's co-pilot Darrel Brown suggested to Bill: "I'll fly as captain if you'll be my co-pilot." The Bold Scot just didn't know how to say no, so away they flew, swooping above Jack's boat as they made a circuit to gain altitude. Jack recognized the plane, of course, and raced furiously back to the airstrip, arriving there just as the skyjackers landed. He threatened to lay charges but soon calmed down. He told me later: "It was bad enough them stealing my plane but I really got mad when they buzzed my boat!"

Prior to the C-46 heavy hauls, Pacific Western introduced two Otters into Portal Camp to carry freight. The pilots were Ron Wells and Pat Carey, both older, experienced pilots. In the long days of early spring, they would be in the air at daylight, around 4 a.m., making the necessary circuit over Stewart to gain altitude. The Pratt & Whitney engines were very noisy and they woke up many citizens but no one complained because they knew the importance of mine freight. However, the following year, there was a bumper crop of babies born in town. The married miners blamed the Otters: "Too early to get up and too late to go back to sleep again!" Bill Ross said each pilot flew more than 200 hours during the month of May that year.

Pat Carey was an old-time pilot, the one who met the challenge of carrying Tony the horse to Tom Mackay Lake in 1946 (see the Eskay Creek chapter.) In his youth, Pat was a wild one but later in life he got a dose of old-time religion. The miners used to say, "In the old days, Pat flew with the control stick in one hand and a bottle of whiskey in the other. Now he flies with the controls in one hand and a bible in the other!"

In 1962, Pat became the part-time lay magistrate in Stewart. Stan Grimson, who was the RCMP constable in town during these years, told me a story about Pat and the boys, Henry and Rocky. They were war vets who liked to re-fight the battles over a bottle of whiskey. One day, they got into an argument that ended with Rocky slapping Henry and breaking a tooth. Henry headed for the police station, where he laid a charge against Rocky despite Stan Grimson's advice to the contrary. When court day arrived, the two of them plus Henry's wife Hilda appeared before Magistrate Pat. They all had their say, including Hilda, who testified that she didn't know what happened because "we were all drunk."

It was time for Pat to pass judgment. He didn't know much about the law and had an arrangement with Stan to give him a signal — his head moving up and down was guilty; side to side was not guilty. So Stan moved his head from side to side.

Pat said: "Rocky, I find you not guilty of assault but I fine you $15 for being an asshole."

Rocky: "But Pat, I don't have $15."

Henry: "That's okay, here's the money, now let's all go to the bar and have a beer."

As my hotel and bar was the social centre of town, the pilots always congregated there, not normally for serious drinking but to enjoy the local characters and to dream up tricks on each other. One winter in particular, quite a few members of the air crews were residing upstairs at the hotel and so the tricks proliferated. Among the best was by Bill Harrison, who borrowed the portable fog horn from Al Lawrence, the port

PACIFIC WESTERN AIRLINE'S C-46 LOADING DYNAMITE FOR THE MINE AT STEWART AIRSTRIP. BILL ROSS, THE OLD BOLD SCOT, IS IN THE DOORWAY.

captain, and hid it in his room at the hotel. When the weather was still bad the next morning, he conned the maid into opening the door where his two pals were still sleeping after a late night. The foghorn worked by pumping up pressure so that when you pressed a button, it would go off with a mighty "oh-oo-yaa." Bill slipped the foghorn beside the beds, pressed the button and headed for the door. As he gave a last glance, the horn went off and two bodies lifted horizontally off their beds, fell back down and hit the floor on a dead run. Fortunately, Bill was long gone or there would have been murder that day!

Northwestern BC and southeastern Alaska are among the toughest places in the world to fly, due to the weather and the mountains. In 1968, Pat Carey was flying his Otter to Portal Camp with a load of oil barrels when a snowstorm hit. When he tried to bank out of it, he hit the mountain right in the middle of a gully but about halfway down a thousand foot cliff. The impact broke off the tip of each wing, which must have slowed the crash somewhat, but Pat was still badly hurt. A helicopter spotted the wreckage and saw that Pat was still alive, but had to head back to the mine to procure a long rope for the rescue operation. Hovering 200 feet above him, they swung the rope in Pat's direction until he was able to tie himself to it. Then they were able to lift him out. The worst of his injuries was a blow to the head so severe that he eventually went blind.

During the busy Granduc years, Pacific Western flew Grumman Goose aircraft between Prince Rupert and Stewart. In 1970, during some bad weather, a pilot took a wrong turn and had to do some tricky manoeuvres to get out a dead end valley. Two of the passengers were top officials of the Granduc parent company. After that scare, they decided there had to be a safer way to travel in bad weather, so the company bought the MV Lumba Lumba, a 90 foot long twin engine cruiser that had been used in the rough Strait of Malacca near Indonesia. They contracted Rivtow Marine to crew and operate the vessel, and it ran nine-hour trips twice weekly from Prince Rupert for a number of years.

In 1974, Captain Vince Brown, the Prince Rupert manager for Rivtow, took the Lumba Lumba on a rescue mission south of Rupert in a terrible storm. The effort was successful and two lives were saved. There is a great account of that effort in the book Dangerous Waters by Keith Keller (Harbour Publishing.)

Through the years, there were a number of fatal aircraft crashes in the area. The first was likely the Stranraer "Skeena Queen" making a mercy flight from Stewart in September 1945, heading to Prince Rupert with a local mother and baby on board. They tragically crashed into the water near Port Simpson with loss of all on board.

But on the positive side, there were also many successful mercy flights undertaken. Often the pilot at the control stick was Paddy McNeil, one of a fine crew of people working with Pacific Western. After some years with the Goose aircraft, Paddy decided to train to fly helicopters instead. Sadly, within two years of beginning this new career, Paddy was killed in a helicopter crash. Just at that time, the District of Stewart was completing the new swimming pool and council chose to honor Paddy by naming the pool after him. The Paddy McNeil Memorial Pool is named in appreciation of Paddy and all the other young men who were killed during development of this treacherous corner of British Columbia.

Bill Ross, who is also a bit of a poet, wrote this for other important members of the air crews.

You have heard of our C-46s
And all of the flying they do
Jim Siddle with nerves of cast iron
And Jack Magel to name only two.

But what of the poor engineers
Who work in the cold night and day
Tuning up C-46s to get Granduc
  freight away.

They're up in the morning early
...if they do get an hour in bed
Still weary from greasing and fixing
And feeling like someone half dead.

And when they do come off duty
And two miles to Hyder go
To have a drink in the evening
And think of our sad tale of woe

Then it's back to the Stewart airstrip
And overalls rigid with grease
To patch up Jack Magel's motor -
He again got it back in one piece.

Hats off to the skilled engineers
It's on him that this air freight depends
It's his work and devotion to duty
That got it all there in the end.

OLD TIME BUSH PILOT DOUG CHAPPEL LOADING OTTER ON WHEEL-SKIS. STEWART STRIP.

Chapter Thirty-Three

THE McLEOD CLAN

FIFTH AVENUE IN 1925; HIGH TIDE ON THE MAIN STREET. KING EDWARD HOTEL IS AT THE REAR, TO THE LEFT.

Our dad, John McLeod and our mother, Catherine, were born on the Isle of Lewis in northwest Scotland. Dad came to Canada at a young age and turned to mining in order to earn a living. However, he had always nursed a typical immigrant's dream — to own a piece of land that would support a family.

Before the railway reached Fort George in northern B.C. in 1910, Dad had staked out a homestead in the area. He went to work constructing a cabin and clearing land with the intention of planting a crop. After he had cleared one acre, he hired a neighbor with a horse and plow to break the ground. That wasn't too difficult as there were only a few inches of soil and below it, coarse gravel. He told us later: "I knew that pota-

THE STEWART TOWN SITE IN 1965 SHOWING THE FIRST GRANDUC HOUSES UNDER CONSTRUCTION. THE KING EDWARD HOTEL IS IN THE FOREGROUND.

BUILDING THE KING EDWARD IN 1910

teaching in Scotland until 1925 when they married. I was born in 1927, my brother Don a year later and our two sisters, Isabelle and Janet, a few years after that.

Like most of his generation, Dad encountered difficult times but he persevered. All through my boyhood, he was a 12-hour-a-day man and on Sunday, his day off, he would be exhausted. In 1940, he bought out his partner in the hotel for $4,000, which he had to borrow. The comparison to the $40,000 that he had paid for the first half in 1923 just goes to show the effect of the Great Depression. When he died in '52, he had paid off the

toes wouldn't grow in potato-sized rocks!" Naturally, he gave up on that homestead which today is exactly the location of the Prince George airport runway.

So he turned back to mining. In 1919, he worked on the B.C. Coast at Alice Arm where he heard of the rich Premier strike and decided to move to Stewart where he lived for the rest of his life, until he passed away in 1952.

Dad mined and prospected in the area, made some money and bought a half-interest in the King Edward Hotel in 1923. Mother had been

loan and had a few dollars to the good. At that time, a close Scottish friend said: "A Scotsman's dream is that when he dies he would not owe anyone a penny and that he have a good suit of clothes to be buried in." If that be true, then Dad died happy.

Like most of the Stewart townspeople, Dad had a nickname. People called him "The Black Prince" partly because of his very black mustache and hair in his younger days. The Prince moniker came about because he was a prince of a fellow — always thoughtful and helpful to his fellow man — and he was certainly that, and more, to me.

Mother's teaching background meant she ruled her children fairly strictly. The strap was available and used if necessary. For the few times I stepped out of line and got it, it did me far more good than harm. Our parents spoke Gaelic but only used it to discuss private things because we couldn't understand. I guarantee there is no better motivation for a child to learn a second language!

Dad always praised Canada over Scotland. He would say how wonderful it was to be able to fish and hunt and own your own land freely compared to the old country where the lords owned the streams, the deer forests, and the crofts. Jail was the reward there for any poor man who dared take even one fish.

THE NEW KING EDWARD IN 1980

Mother was more patriotic to Scotland. She never did lose her accent, insisting that she spoke the proper "King's English" and that the rest of us sounded funny! And she had a grand sense of humor. One of her favorite stories was about going to a new doctor for a checkup after moving to Vancouver. The doctor was questioning her:

"Have you ever smoked, Mrs. McLeod?"

"No, never had a cigarette in my life."

"Do you take a drink, Mrs. McLeod?"

"Aye, I do, providing the companion I'm with pays for it!"

Dad came from a family of thirteen children of whom eleven immigrated to either Canada or the United States. The two who remained in Scotland were Aunt Annie and Aunt Margaret, both of whom never married and were always thought of fondly by us. When Aunt Annie visited Vancouver in 1972, we went to see her. I happened to be sitting beside her, with the television playing across the room. To make conversation, I asked "Annie, do you have TV at home in Scotland?"

"Yes, Ian, we have everything you have here — and more!" That told me off!

Brother Don and his wife Christa visited Annie in Scotland in the 1970s. One Saturday

BROTHERS IAN AND DON MCLEOD, 1933.
BEST FRIENDS AND PARTNERS FOR 75 YEARS.

night, Don dropped into the local bar to sample the "water of life," as the Scots called it! When he arose on Sunday morning, he was feeling a bit rough so he asked Annie if she had a wee dram in the house. She produced a bottle for him, but first drew the blinds so the neighbours couldn't see the Sabbath being broken!

After I left the army, Don and I worked and travelled together. We mined and packed horses and had many other adventures. When Dad died in 1952, I quit mining and started working on the old hotel. I only came to the end of that project some 49 years later, when I sold the business!

Don continued mining, eventually reaching the position of manager at Premier in 1964 where he had started as a mucker a decade earlier. In 1976, he started Northair Mines and as president brought the Brandywine Falls mine near Squamish, B.C., into production. It operated for six years at a profit. In 1978, he formed Scottie Gold Mines and brought into production the Morris Summit property, located at Summit Lake beside the Salmon Glacier. It was unprofitable and closed in 1985. Next he formed Newhawk Mines and started exploration on a property at Brucejack Lake, northwest of Stewart (see "Murder in the Mine" chapter.) Don and Christa have two children, Catherine and Bruce, both of whom are also successful miners.

My wife Vivian and I also have two children, Nancy and Robert. Nancy works in the world of commerce, but Robert is a geologist so together Don and I have three third-generation miners in the family.

My sisters left Stewart as soon as they were graduated from high school, as did most of the girls we grew up with. There was not much for a female in that kind of country!

The King Eddy

When I started at the hotel, it was an old building with no heating system, just two pot-bellied wood stoves, one in the lobby and one in the bar. The rooms froze in the winter but then there weren't generally many travelers at that time of year either! One old-timer told me about an extremely cold winter when a Swede miner checked into a room. By 5 a.m., the miner was so cold he dressed and went down to the lobby to warm up beside the stove. Unbeknownst to our Swedish customer, the local milkman had just gone upstairs to the washroom and came down right after the Swede, who was rubbing his hands over the stove. He looked up to see the milkman coming down the stairs with icicles hanging from his beard. "Yesus, vat room were YOU in?" asked the astonished Swede.

Within a year, I had central heat installed and started renovation and expanding. Granduc mine was bringing men in and road construction to Cassiar had begun. Business at the hotel was going well when one cold February night in 1971, the place caught fire and burned to the ground. Thirty-seven sleeping customers thankfully escaped without injury, but the building burned down with many valuable old pictures and artifacts inside.

It was, however, insured so I immediately ordered a twenty-room modular motel that was operational within a month. Then with two good men, Dave Meneghello and Max Ryder, we started building the present "King Eddy". Within a year, business was back in full swing. When I recall those years, I realize how blessed I was with an especially fine crew of 30 to 40 employees, many of whom spent a lifetime working with me.

In the early days, my King Eddy was the social centre of town where the barber set up shop in the lobby and everyone visited the bar. Drinking was a secondary reason to come to the bar — socializing and "BS'ing" were the first, and sometimes listening to the local singing troubadour, Max Ryder, as he strummed his guitar and sang for us. A pal repeated to me what he had heard from a stranger to town. "I've never seen a bar like that, with twenty people sitting around, reading and talking; nobody drinking and at the corner table is the owner, cutting up a halibut he just caught, wrapping the pieces in newspaper and giving it all away to the old-timers."

Since Highway 37A connected Stewart to the rest of the world, we have experienced a large influx of tourists. In my last decade at the hotel, we catered to many coach tours, many of which were from Europe. One day we were expecting a German group but that afternoon their guide phoned me from Hazelton. He told us they would arrive around dinner time and asked if we could cook up some salmon the group had just purchased from the natives in Hazelton.

AUTHOR'S HOME, STEWART 1978

HAMMY HAMILTON

"Are the fish clean?" I asked.

"Yes, very clean," he replied, so I assured him that we would cook them.

There were 20 or so pink salmon, each about three or four pounds, one for each in the group. They were clean — but not cleaned, and they were also spawners in extreme deterioration — something a bear might eat but not a local person. I asked my chef, Steve Kirkbride, what to do.

"Don't worry, Boss, I'll clean them, wrap them in foil, pour the lemon juice on and bake them."

I went home that night fully expecting phone calls of complaint, but nothing happened. The next day, I asked Steve how it had gone.

"When they started eating, I kept checking with the waitress and she said that they were all cleaning their plates. Then they asked her to bring the chef into the dining room. Well, I was ready to run for it if I had to! But they all stood up and gave me a standing ovation. Wow, I didn't expect that...." Neither did I!

One of my bartenders in the early days was 'Hammy' Hamilton. He was a fine jolly fellow whom everyone liked but my problem was that he liked the customers so much, he was always buying them beers! Not even with my money, but with his own. When he bought someone a drink, he signed a tab for me to take off his pay cheque. One payday, I discovered that he had absolutely nothing left for himself, so I put my foot down and forbade any more of that practice.

Sometime after this, a friend described to me how he had arrived at the bar one morning to find Hammy crying

"What's wrong?" he asked my bartender.

"Ian just fired me."

"Why did he do that?"

"I don't know but it's not payday and he just came in and gave me this cheque."

"Let me see it," said our mutual friend, who examined it and told Hammy: "Look, this is your holiday pay for last year. You're not fired."

Whereupon Hammy brightened up immediately — "Good! Let's have a beer!"

I always had a few old-timers staying upstairs at no charge because "the dole" didn't stretch enough to pay for a room. But I could always get a bit of work out of them. One such tenant was Dennis Delaney who was known as Captain

Morgan thanks to his favorite brand of rum by that name.

In the early days, the bars were closed on Sundays and I would spend that day fishing, which I loved. On my Sundays off, I would give Captain Morgan the key to the bar and he would swamp the place out with mop and pail. But because of his addiction to drink, I would always count the beer beforehand (in these days we only sold beer.) This went well for a few weeks but one Monday morning, I arrived at the bar to find the mop and pail sitting in the middle of the floor, the job only half done. I really didn't have to count the beer to know what happened but, as I suspected, there were nine bottles missing.

Later that day Captain Morgan slunk into the bar, hung over and looking sheepish.

"You old coot, you didn't finish the job and you stole nine bottles of beer," I told him. To which he replied: "Well, how about giving me three more and make it an even dozen."

Stewart became a thriving and vibrant community when Granduc was in full production.

In 1978, though, Granduc closed the mine and hundreds of people left town. The company boarded up the vacant houses and local people were depressed. Somewhere out of the blue came the news that the movie "Bear Island" would be filmed in Stewart during the winter of 1978/79. Many of the townspeople were offered jobs, the boards came off the housing and more than a hundred film workers arrived in town. My hotel was filled with the workers and things were rosy again. Money flowed and by the time film production was finished, I had enough to pay off the mortgage. Like my dad before me, I could now die a happy Scotsman! All I needed was a good suit of clothes....

BILL CRAWFORD AND JACK MCLEOD, 1940.

The filming presented many interesting incidents in Stewart but I remember one more vividly than the rest. I was sleeping early one morning about 5 a.m. when the hotel clerk phoned me to say that the waitress hadn't shown up for work but the movie crew was up and waiting for breakfast. I judged that the best solution was for me to run down quickly and substitute for the waitress, which I did. But after I'd spilled some coffee and

DON McLEOD, 1995.

dropped toast on the floor in front of the star Vanessa Redgrave and a couple of cute starlets, they told me to sit down and have breakfast — they would waitress! And they did it very well, commenting after that if the movie business didn't work out, they could always take up hospitality instead!

My name, Ian, is Gaelic for John and in the 1930s, it was a fairly rare name although there seems to be many of us around now. I knew many Scots whose first name was John including my father, who was called Jack, but I was the only "Ian" around in those days. As a youth, I was called "Evie" by my chums and the old immigrants had a wealth of pronouncements including Eno, Eenie and Ieano. When I started mining, I was sent to help an old Swede who asked my name. After a few attempts at pronunciation, he said: "I call you Yack yust like you daddy." So I was also Jack to some of the miners!

It was with much regret that I sold the hotel in 2001 but the old body was not really up to the chore anymore. Vivian and I now live in Kelowna but still keep our Stewart house and return to the north during the summer months. Kelowna is nice but I miss the north, except for two things — - snow shovelling and mosquitoes.

RED CLIFF MINE CAMP, 1910. THIS WAS AS FAR AS THE PGSL RAILWAY WAS BUILT.

WRANGLERS BREAKING MUSTANGS TO SADDLE, 1925. MT. DOLLY BEHIND.

SIX HORSE TEAM AT CASCADE CREEK BRIDGE WITH LOAD FOR BIG MISSOURI, 1928.

Chapter Thirty-Four

NICKNAMES & CHARACTERS

In the 1930s and '40s, as I was growing up in Stewart, I was always intrigued by the strange and innovative nicknames attached to the people who lived there. Some were understandable as a lot of the people hailed from Europe with unpronounceable surnames but some of the nicknames were downright cruel.

There were many Mikes around and in the forefront of my memory are Concrete Mike and Greasy Mike, the proprietors of the two restaurants in town and both passable cooks and bakers. Concrete Mike, a Croatian, rented the hotel restaurant so I knew him well. On one occasion, an upcoming municipal election was pitting Jack Doyle against Bill Tooth, the long-time incumbent mayor. Doyle was a customer of Concrete's, taking all his meals in the restaurant. On election day, Mike naturally voted for Doyle but Doyle lost by a lot of votes. He took it very hard and stopped coming downtown, which also meant he wasn't eating at Mike's. Mike said to me, "Funny country, funny country, vote for, lose the business, no vote for, lose the business...."

Mike had a small safe in the kitchen where he kept his cash and papers. But he kept the safe combination pinned to the wall above the safe at all times. Luckily, there were no break-and-enters in those days; fire was the biggest hazard and that's why Mike used the safe.

Other Mikes included: One-eyed Mike, Garbage Mike (the garbage man at Premier), Georgia River Mike and B.C. Mike.

The cruelest nickname was "Broken Ass Shorty," although Shorty never let on that he minded. He was born Nathaniel "Nat" Ernst in Lunenburg, Nova Scotia and went to sea as a young man on a square rigger ship. In the middle of the ocean, he fell off the rigging and broke his hip. By the time they docked and were able to get to a doctor, the hip had reset itself and couldn't be repaired. Shorty was thus crippled for the rest of his life and came out west where he took up camp cooking. By the 1930s, he had retired and was living in a small cabin in Stewart. Our mother always taught us to be pleasant and respectful to others so Mom was horrified when my six year-old sister Isabel related to us one day that she had met Shorty on the street and said, "Hello, Mr. Broken Bum."

Shorty had saved a few dollars so after he sold his cabin, he was able to return to lovely Lunenburg for his final years. When I visited the museum there a few years ago, I recognized that the Ernst family had been very prominent in shipbuilding and seafaring.

* * *

Hungry Harry was another memorable character. He was the cook at the Big Missouri Mine and would tell folks, "When the new men come to camp they are very nice for awhile but once they get the wrinkles out of their bellies, they start to complain about the grub. But I know how to fix them. Find out what they don't like and feed them plenty of it!" When Harry left the mine, the boys put up a sign in the cookhouse that read: 'Hungry Harry, gone, but not forgotten.'

* * *

Fred Hasil was an Englishman known locally as Lord Plushbottom. When he first came into the district, he dressed to the nines and seemed to have a lot of money. However, by the time I knew him, he was living in a shack near the Bear River bridge and was just getting by. But he was always a proper gentleman. Fred loved cats and when I was a boy, he had about twenty of them. One time, when the female cat at the McLeod house got pregnant for the third time, Dad decided to get rid of it. I had a .22 rifle and thought that I was a great hunter so Dad instructed me to take the cat into the woods and shoot it. I put the cat in a box, hoisted my rifle and headed for a secluded spot in the bush. But when I took the cat out of the box, she began to purr. I just couldn't summon the courage to kill her, so I packed her up again, hiked up to Plushbottom's cabin, and released her. Weeks later, Dad hinted he knew what I had done, but no more was said about the matter.

* * *

Lorne Dempsy described himself as a tramp miner, as did many others. These men would work the camps for many months, save their money, and then go to town with full-time parties and boozing until the money was gone. In 1948, my brother Don and I were in Prince Rupert looking for work. A miner we knew told us that the Taku mine was hiring and that we could sign on at the P. Burns Meat Market warehouse office. During a conversation with the clerk there, he also showed us the telegram he had received from the mine. I have never forgotten it.

"Please forward 3 hind quarters beef, 4 sides pork, 50 doz. eggs, 100 lbs bacon, 3 miners, 2 muckers, 1 timberman, 3 laborers."

We hired on as muckers and were given tickets to Juneau, Alaska via the CPR's "Princess Norah" and from Juneau, we would fly into the mine that was located on the Canadian side of the border. Dempsy got on the plane with us and when we arrived at the Taku strip, there was a bus to drive us several miles to the camp. As we loaded our luggage onto the bus, the driver, who knew Dempsy well, said: "Dempsy, where is your packsack?"

"Oh, my head is sticking out of it!"

He must have had a particularly wild time in town, to spend everything including the packsack, and it reminded me of a little ditty the miners and loggers used to sing about their lady loves.

"Goodbye John
Your stake is gone
Your packsack's at the door."

* * *

Don and I were once mucking (shoveling the ore into cars) behind an Irish miner named Jimmie Flannery. Jimmie would drill and blast on day shift, no work would be done on afternoons in order for the dynamite smoke to lift, and then we would go to work on the midnight shift. More often than not, as we were leaving the portal at seven in the morning, we'd meet Jimmie heading in. He'd ask us: "How did the round break last night, boys?"

We'd likely respond: "Not very good, Jimmie. The lifters (floor holes) didn't break well at all."

"Ah, boys, t'is a hard way to make whiskey."

* * *

Dr. Mike Guard was the town physician during the Granduc heyday. He was a fine doctor with a great sense of humor. His family lived in Vancouver and one Christmas season he took time off to join them for the holidays. Getting away meant driving 150 miles to Terrace and flying from there to Vancouver. When he left Stewart, there was a bad snowstorm in the northwest and at Terrace he found that a number of flights had been cancelled. The airport was full of people anxious to get on the first available flight south. He saw on the monitor that his flight was cancelled, too, but that another plane was just landing. At the same time he spotted the Stewart RCMP constable in the lineup and went over to chat with him. As Mike recounted his bad luck to the constable, the police officer had an idea.

"Mike, do you badly want to go to Vancouver?"

"Yes, of course."

"Would you go in handcuffs?"

"Well, sure."

"Hold out your hands," said the officer, and he put Mike in cuffs. When they got to the airline agent, the policeman told the agent that he had a problem: just one seat had been reserved for himself, but now he had a prisoner. The agent said he would bump a passenger to make room for the prisoner. The policeman said to the agent: "Thanks. He's a mental patient and he thinks he's a doctor."

Mike made it home for Christmas.

* * *

George Wright was a Newmont Company prospector in the 1960s who was working out of a fly-in camp on Todd Creek. He flew into Stewart to pick up supplies and just as he was preparing to drive to the airstrip, I came along and stopped to chat. George spotted two nice salmon in the back of my pick-up truck; I'd caught them only that morning and they weren't even cleaned yet. I offered him one, put it into a plastic garbage bag, and away he went. Apparently, when George arrived in camp later that day, he stored the garbage bag in his tent and that evening took up his fishing rod, informing the Chinese cook that he was going fishing at the creek. The cook told him, "George, no fish in creek, too fast water."

"That's okay, Charlie, I'll have a try anyhow."

George snuck back to his tent, took the salmon from the bag, poured fresh water on it, and took it to the cook. "Charlie, you clean my fish now and we'll have it for supper tomorrow."

Every evening for the next week the cook was at the creek trying to fish, but never with any luck.

* * *

George Kustus was known as George the Greek. He ran the stage line to Premier Mine for most of the 1930s. The first service to the mine was by horse-drawn open sleighs in the winter and rubber-tired vehicles in the summer. George had a small 'cat' that towed a caboose with a stove in it so the passengers travelled quite comfortably. In 1934, the government procured a big plow with Caterpillar power and before long, Premier was accessible year-round to motor vehicles. Some families moved to Stewart and George drove the men to work every day. When the mine closed in the '50s, George moved to Prince Rupert where he was tragically killed by a car while crossing the street.

* * *

John Angus MacDonald was a veteran of both world wars and a very stubborn Scot who loved Scotch, the "water of life." In the early days, the Provincial police were very strict about drunks. John Angus was picked up a number of times for his drunkenness and finally 'Hoot Man' gave him six months in Oakalla Jail.

* * *

'Hoot Man' Billy Orr was the government agent and lay magistrate from the '30s to the '50s. He is not fondly remembered, as he was notorious for never finding people 'not guilty'.

When John Angus was released, he swore he would someday pour a bottle of whiskey over Hoot Man's grave. He was eventually in a position to carry out his promise, but he said: "I kept my promise to pour a bottle of Scotch over his grave but I filtered it through my kidneys first."

The police were also strict on natives, and I remember when three Gitsan natives were each given six months for possession of a bottle of American whiskey. When the RCMP took over from the Provincial police, a more common sense approach was used and the fellows with a snoot-full were generally just provided with an immediate ride home.

In 1938, Fred Larson came to town, rented an office and put out his shingle as a chiropractor. His clients seemed to think he was okay and many returned for his manipulations. But it turned out that someone in town had worked alongside Fred in a mine in southern B.C., and they were suspicious. When the constable investigated, the only certification Fred could produce was a blaster's permit. Thirty days was the sentence, far more lenient that what was usually levied to drunks and Indians!

* * *

Ski-pole Hazel was a retired lady from the houses of ill repute. Her hip had never healed properly after a fall and so she needed an aid to help her walk. A ski pole worked just fine and because there was snow and ice at least six months of the year, she decided she might as well use it year-round. Hazel had a small house and even in her old age, she liked to party it up with the boys. At one of the parties, one of the fellows spotted a chamber pot under Hazel's bed and a bottle of Epson Salts on the counter. So when Hazel wasn't looking, he poured some Epson Salts into the chamber pot. Next day, he happened to meet Hazel hightailing it for the hospital.

"What's the problem, Hazel?"

"I'm peeing white foam and I've gotta get the doctor to check me over."

The fellow took pity on Hazel and told her what he had done. His reward was a good crack on the head with the ski pole.

* * *

John McKenna operated what was called the drugstore from 1935 to 1960, when he retired. John sold over-the-counter medication, papers, books, candies, etcetera, but no pharmaceuticals. Although of Irish descent, he was very close with his money and I often heard him say "Look after the pennies and the dollars will look after themselves."

John didn't drink often, but when he started, it usually resulted in a three or four day bender. The boys used to say, "When McKenna is sober, he's tight, but when he's tight, he's very loose." John would visit every table in the bar, buying drinks and leaving his change on every table. His wife Dorothy didn't like these antics and sometimes would come looking for him, but John seemed to have a radar for Dorothy. When he'd head out the back door of the bar, we'd watch and, sure enough, a few minutes later Dorothy would come in the front door.

John had a little coffee bar in his store and we would go there at the morning break. One time, I was sitting with my partners Angelo and Ramo. (The 'f' word was very common amongst the miners but I prefer to write it as "trucking.")

Angelo said: "There's going to be a trucking big mine here someday, boys."

Ramo kicked Angelo under the table and said "Watch your language, there are ladies in here."

"What the trucking hell did I say now?"

* * *

Harry Hanhart was an easy-going Swiss fellow who fell in love with the Portland Canal and adjusted his life so he could spend as much time as possible on the water. Harry was working for me as a carpenter and one morning he didn't show up for the job. In the afternoon, he came around to quit.

"What's wrong?" I asked.

"Well, I've had some bad news and some good news this morning. The bad news is that my father died in Switzerland and the good news is that I've inherited a quarter of a million dollars. The first thing I want to do is go to Vancouver and buy a decent boat."

So Harry went to the city, found a boat he liked, became instant friends with George the boat salesman, and paid him $25,000 for the boat. Certain upgrading had to be done so Harry came home for several weeks and then returned to pick up his new boat. But he found out that George had skipped town with Harry's money and as the boat had been held by the dealership on consignment, the original owner had not been paid. So Harry paid out another $25,000 as he really liked the boat.

It turned out that George was an alcoholic who blew the whole wad on a big drunk in California. But that wasn't the end of their association... Harry next bought a building, renovated it and started a sporting good store. Who should he hire as his manager but George, who had sobered up and returned. When I asked Harry why he hired George, he replied: "When a person is such a good salesman that he can sell me a boat he didn't even own, I want that salesman working for me. But also I might be able to deduct some of what he owes me from his wages." But that didn't work out because George went on another bender and Harry had to fire him.

Fishing was also a favorite pastime of mine, and when I was in the market for a new trolling rod and reel, I sized up various makes and models in Vancouver, priced them at about $200 there, and then ordered them through Harry's store in Stewart. A couple of weeks later, I stopped in at his store to pick up the order, and he said I owed him something like $100.

"Harry, I priced these items in Vancouver and the cost was at least double that," I told him.

"Well, you know, a fellow doesn't feel good to make a profit from a friend," he answered. Harry had no enemies.

Harry spent many years fishing on the canal and he was never seen without his captain's hat. We were very grateful for his presence on the water, especially on one occasion when we found ourselves stranded without gas about four miles out of Stewart and Harry came to our rescue.

* * *

Ted Setness was an old Norwegian miner, broken down and crippled but with a dry sense of humor. During the grand opening of the hotel's new cocktail bar, the place was so crowded the bartenders were run off their feet. Ted came in the door so I handed him an apron and asked him to clean tables and pick up empty glasses.

TED SETNESS AND TOMMY PALMER AT MEZIADEN LAKE. NOTE THE SAW UNDER THE WINDOW TO DISCOURAGE BEARS.

He did so well that I hired him as a daytime bartender because things were quieter on the day shift.

The weekly steamship arrived on Saturday with round-trip tourists and while most of them preferred to visit the bars in Hyder, sometimes a few came to Stewart. Al Lawrence was sitting at the bar talking to Ted one such Saturday when four tourists came in. Ted took their order at the table and came back to the bar to mix four whiskey sours. Al said to him: "Ted, these aren't whiskey sours, they're Tom Collins." But the recipe for Tom Collins was all Ted knew.

"Oh, well," he replied to Al, "they don't know the difference anyway."

CHEF BOB BARWISE WITH TWO FINE SALMON READY FOR THE BARBECUE.

Ted took the drinks to the table and collected the money. Al said the tourists each took one sip, got up and walked out. Apparently they did know the difference.

* * *

Another good friend, Bob Barwise, spent the first seventeen years of his working career in the parts department of the city of Vancouver. But the north called to him, and when he got to Stewart he thought he had arrived in heaven. Bob's personality fit perfectly with the old-timers. Once he was looking for work and the only job available was as a cook. He had never cooked before but thought he could handle the job if he had some recipes. By then Bob knew which wives in Stewart were considered the best cooks, so he collected all their favorite recipes. After that first job, he never looked back and he was in great demand. Bob cooked in my hotel for years as did his son Doug. Although Bob liked the bottle a lot, he never missed a shift or was late for work.

* * *

Joe was another old-timer who lived around the corner from my house. He had a vegetable garden where he grew potatoes, turnips, onions, etc. One summer I noticed a lot of black bears wandering near Joe's garden but didn't think much of it. However, Doug Miskow, the Canadian Customs officer, related a story about Joe coming through Customs one day on his way back from Hyder. When Doug opened up the door of his van to ask Joe if he'd anything to declare, an overwhelming smell just about knocked him off his feet. It was a couple of dozen rotten salmon from Fish Creek in the back of the van. Joe was apparently using them as fertilizer — - they worked great on the vegetables because his turnips were always as large as footballs, but obviously the bears liked it, too, even if they had to dig for them!

* * *

Bonus Nick was a miner who worked at Premier in the early forties. He was small in stature but strong and very ambitious. The mines paid an incentive bonus for individual production over a certain level, and Nick worked very hard to earn that bonus — hence his nickname. When Nick decided to go into business he built a restaurant and rooming house on Fifth Avenue. Like so many Stewart businesses, it was lost to fire.

The village of Stewart hosted a public meeting after one of the numerous mine closures to figure out what we could do to keep the town from dying again. Nick stood up and made his only public speech ever.

"You people talk about another mine but you better forget it," he said abruptly. "Very little ore around here, just a little bit here and a little bit there, just like the bear shit on the hill. Look to timber for the future, you can see the trees and count them." With a fairly prosperous timber industry in the area now, I can see that Nick was right.

* * *

Mike Zuluk was a retired Polish-German miner. One time during the sixties, my in-laws travelled from their home in Edmonton to visit us. They took the train to Prince Rupert and boarded the Northland steamship for Stewart. As the ship was pulling out, Mike and my father-in-law, Fred Seifert, started a conversation.

BONUS NICK ON LEFT, WITH JOHN LUNUK.

When Mike heard that Fred was coming to visit his daughter and son-in-law, Ian MacLeod, Mike declared loudly: "ZWANZIG, ZWANZIG, Ian is my best friend!" Zwanzig means "twenty" in German but although Fred understood a little German, he couldn't quite figure it out. Mike then invited Fred and his wife Natalie to the stateroom for a drink. When Natalie described the situation later, she was still in shock for Mike opened up a suitcase that was utterly packed with every sort of liquor possible. Stateroom parties were quite common in those days but my in-laws were not the party type — one drink at a time was all they wanted! Upon their recounting of the story, I also had to explain that Mike called me "Zwanzig" because whenever he was short of money, I would loan him a $20 bill.

* * *

The Fiva brothers, Gunn and Inge, were bachelor Norwegian prospectors who lived at Alice Arm, a small mining community south of Stewart. When my good friend Bob Switzer was filming in the area for CBC-TV, I often accompanied him as a guide. Once, we flew to Alice Arm and Bob was interviewing the Fivas about mining and prospecting. He asked Gunn, "Mr. Fiva, were you ever married?"

"No, no, but I had a girl friend once," he said, and pointing at me, added: "Ya, and he knows her, too."

"And what happened?" Bob asked.

"Vell, I asked her to marry me and she said no. A few years later, I staked the big B.C. Moly mine and made a lot of money. So then she asked me to marry her. And I said no!"

THE HOTEL AT ALICE ARM WHERE THE SHOOTING MATCH BETWEEN HAROLD AND SHORT CIRCUIT TOOK PLACE

Gunn told a tale of a shooting war at the old hotel in Alice Arm, owned by Harold Lund. Alex Mowatt, an electrician known as Short Circuit, was drinking with Harold one day when they got into an argument. It got so hot and heavy that they decided to shoot it out. Harold was on the upper floor of the hotel and Short Circuit was in the yard, behind an old vehicle. Bang, bang, bang. Then Short Circuit waved his arm at Harold and yelled, "Stop shooting. I'm out of ammunition."

"I'm not." Bang.

Fortunately, no-one was hurt and the party resumed.

Harold was married to Kay. At one of the miners' parties, she was bragging about her youth when she was a beauty queen and married to an Italian Count. She told of them travelling to Rome once, where at a party she met and danced with Hitler, Mussolini and King Emanuel. The King fell in love with her and since the royal prerogative could not be refused by the Count, Kay had to go to bed with the King.

One of the miners commented: "Well, Kay, you must have got a royal trucking that night."

* * *

Oliver 'Cabbage' Smith was the local market gardener, and his facility was located at the north end of Brightwell Street past the present elementary school. He operated a fair-sized greenhouse and grew and sold potatoes, turnips, carrots and, of course, cabbages.

* * *

Eva 'Momma' Litwin was born in Ukraine and farmed in Saskatchewan with her husband until he passed away. She then moved to Stewart and worked for me in the hotel for more than 30 years. Everyone called Eva "Momma". She was my head dishwasher and in her spare time she cooked perogies, cabbage rolls, doughnuts and cakes that were loved by all. Eva called me "Boss" and my wife Vivian, "Mrs. Boss." Vivian is not very bossy and one day, while having tea with Eva, the conversation went like this:
- "You have a nice garden, Mrs. Boss."

- "You know, Eva, that my name is Vivian."

- "Yes, I know that, Mrs. Boss."

End of conversation.

Eva had a nice home in Stewart with a large garden. By her mid-eighties she was still working for me and although some people said that I should retire her, I knew that such a move would break her heart. One afternoon, after her shift at the hotel, she went into her garden to do some hoeing. She must have felt some discomfort because she sat down in a nearby chair with the hoe in her hands, and just passed away. We miss her still.

Because Eva called me 'Boss' many of my staff picked it up, too. After a successful venture or two, they started to address me on memos as "BO$$" as they guessed I was getting rich. When I sold the hotel in 2001, my 30 fine employees gave me a farewell plaque addressed to – you guessed it — BO$$!

* * *

Jimmy Cullen was the editor of the Portland Canal Miner in 1910. Mining news was pretty slow but in the fall of that year, Jimmy printed a story about a big gold strike, a virtual mountain of gold at the head of Bitter Creek. Outside papers as far away as London reprinted his article, and a mini-staking rush ensued. Jimmy later admitted that the big strike was only in his imagination but business had perked up in town for awhile. Ironically, in 1992, a big strike was made on Red Mountain at the head of Bitter Creek. Lac Minerals Co. did a lot of work at the site and although it is currently idle, they claim there remains more

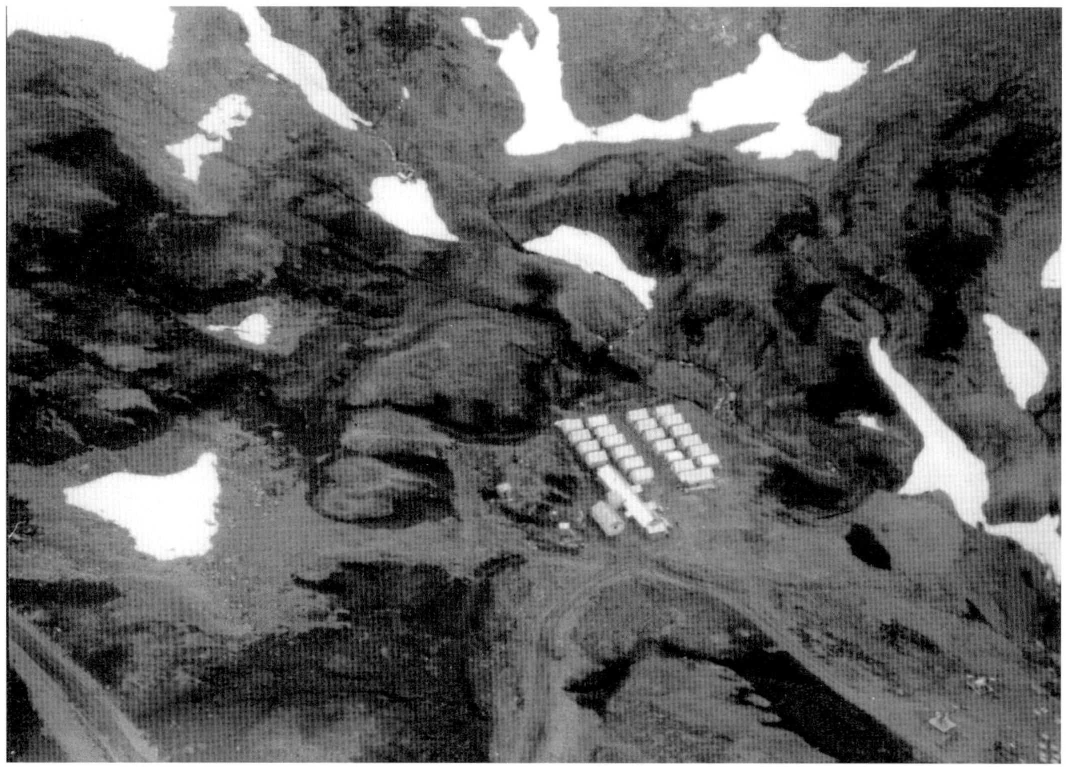

RED MOUNTAIN GOLD CAMP AT HEAD OF BITTER CREEK, 1995. THE DEPOSIT WAS DISCOVERED IN 1989 AND EXTENSIVE UNDERGROUND WORK HAS BEEN DONE ON IT. ALL ACCESS, INCLUDING HEAVY EQUIPMENT, WAS BROUGHT IN BY HELICOPTER.

than a half million ounces of gold blocked out. Perhaps Jimmy's dream will one day come true!

* * *

Hap Kearns came to Stewart in the 1940s and stayed for the rest of his life. Hap was a master of all trades — name it and he could do it. He was electrician, refrigeration mechanic, prospector and undertaker. He worked often for me, wiring the new hotel and house, and when I had an electrical problems, I would always call on Hap, our "house electrician!" If the problem wasn't urgent, he sometimes procrastinated and that was the case with an emergency light in my bar. The light featured a little red bulb that was on when the unit was charged and working but it hadn't worked for quite awhile. One afternoon, I was sitting at the bar facing the light and noticed that it was on again! "Hap must have finally fixed the unit," I thought. Just minutes later, Hap's son Rick came in the door and told me that his dad had just passed away of a heart attack while shoveling snow. I looked at the bulb — it was out and never did light up again. I've often wondered if that's how Hap had said goodbye to me.

* * *

Dave Ashwood was nicknamed "the running shoe prospector." He wore running shoes rather than the usual heavy leather boots for his footwear, in a time when adults did not normally wear the youthful shoe, and thus he was nicknamed. Dave told me he grew tired of wearing the heavy boots that always seemed to be full of water from crossing streams, etc. So he adopted running shoes, cut holes in the lower sides and water would run right out after the stream was crossed!

* * *

Marie Cedar was of Scandinavian origin and probably got her nickname of "Snoose Marie" from chewing snuff, a habit that was common to most of the Swedish miners. She was well known in Stewart, and Sam Kirkpatrick's journal reports that she passed away in February 1951.

One of the old-time miners told me about Marie's trip with Eli Watlin. In 1928, Eli had sold his Prosperity claims to the Premier Company, and he was flush with cash. He invited Marie to accompany him to visit Sweden and visit old friends and relatives. The trip was by train across Canada and then from Halifax to Liverpool on one of the Cunard liners. Now Eli was a big, well-built man at six feet four inches while Marie was a five footer. Eli bought a suit off the rack and my friend, who happened to be on the same ship, said the pants were several inches above the ankle and the sleeves, close to the elbow. This odd couple was the main topic of conversation, he said, especially when the other passengers took note of the big wads of snuff loaded under the lower lips of Marie and Eli. They got rid of their snuff overboard when it was time for the before-dinner drink and then loaded up with Copenhagen again after dinner!

Eli was a prospector and a contract back-packer, the latter meaning he would carry freight on his back and get paid by the weight. These contractors worked on the mining trails that were too steep and rough for horses. Eli was known as a "hundred pound man," which meant that he could carry that amount of payload all day. But work like that had a price, and when I knew him in the early '40s, he was badly bent over and could hardly walk.

* * *

Alan Carolin was a first war veteran who owned a general store and the movie theatre in Stewart. By 1940, with the southward exodus of people to support the war effort, business was so slow that Alan closed his businesses and went to work as a diamond driller. He owned a one-room log cabin on the west bank of Rainey Creek and 500 feet north of the Hyder road across an open field. Alan liked to drink beer when he wasn't working, and he was also a voracious reader. When I would bring in the daily newspapers from Vancouver and Prince Rupert for my bar customers, Alan would read them from cover to cover. However, he didn't care for company and he always drank alone. I was bartending on one quiet afternoon when he dropped in and ordered a beer. He told me, "Bring one for yourself, too, I want to tell you something."

This is the story he told me. "I was born a Catholic but have never had anything to do with the church since I was a boy. I knew they always kept their eye on us so when I was sitting at my cabin table today and saw a priest come off the highway and head for my cabin, I knew that he was coming to see me. I didn't want to talk to him but I had a problem. I didn't have an inside latch for the door and no cover on the window. But I had a small closet with a door so I backed into it and closed the door. It was so small I was standing as if to attention. There were several knocks on the outside door and then I heard it open and then footsteps as the priest walked past the table and bed and paused in front of the closet. The door swung open and there I was, standing at attention and dumb struck. The priest looked shocked, too. Finally I stuttered something like, get out of my home, and he fled out the front door and down the path. (pause) Ian, bring me another beer!"

* * *

Al Lawrence was a close friend of mine with a good sense of humor. In the days of the Northland Navigation steamships, Al was the wharfinger at the Stewart dock. We called him the port captain. To set the stage for this story, it is important to know some terminology from the local mines: they typically worked three eight-hour shifts every day. If you were on a team that drove a tunnel, you called the men on the other shift your "cross-shift." In town, if there happened to be a romance between a miner and a woman married to someone else, they called that "cross-shifting" because if the husband was on a different shift, the lover was relatively safe. There was a miner I'll call Romeo and another miner's wife, a young blonde who I will call Juliet. Rumor had it that Romeo had made it big on the stock market and that Juliet became interested in him — or his money. Romeo and Juliet decided to elope but Romeo didn't want to tell his own wife so he snuck some dress clothes out of the house and on the day the steamship was to arrive, he dressed in his mining clothes, including big knee-high boots that were called muckers. Just before the ship left for Vancouver, Romeo showed up in Al's wharfinger office and quickly changed to his dress clothes, leaving his work clothes and boots in Al's office. After the ship had left, Al took the muckers and spiked them to the wall, drawing an arrow beside them that pointed to Vancouver with the words beside them "He went thataway." We learned later than when Romeo's money ran out, the blonde also left for greener pastures!

When the municipality built the ice arena, Al was hired to run the facility. He loved sports, knew hockey well, and spent all his spare time coaching and teaching. When Al died of cancer in 1977, my village council honored him by naming the arena the "Alan Lawrence Memorial Arena."

* * *

John Meneghello, an Italian immigrant, was a carpenter and a first-class worker. During the 1930s, he owned the impressive 50-room Empress Hotel in Stewart. John did much of the work on the Empress himself but at some point, he decided it was not suitable so he build the Letizia Hotel on Fifth Avenue. In 1961, he sold the Letizia to Claude Bordeau but shortly after that, the hotel burned down. It was a cold windy February and the flames were blown onto the Government Liquor Store across the street. Soon that building was ablaze, too. I served on the fire department at the time and we had limited water to get the fires under control, so we borrowed a large snow blower and with acres of snow to blow (it lay on the ground about five feet deep that night) we soon put the fire out. The roof was gone from the building and all that was showing was a big pile of snow. But under the snow was a treasure — thousands of cases of beer and hundreds of cases of liquor. Both young and old who had a taste for such refreshment had a hey-day. Get a shovel, dig a little, then take home the loot for a party. By the time the snow melted in May, there was a much larger count of alcoholics in town!

John's son David was a school friend and one of my best friends. David spent his life in Stewart and was quite successful. He worked with me building the new King Eddy as well as on my new home. In the fall of 1959, David was working on the Stewart-Cassiar road construction when he and a companion were caught in an avalanche of snow and water. The companion was swept away and died, but David got a grip on a sapling. His ankle was pinned and broken, and he was up to his waist in freezing water, but he held on for hours before rescuers arrived.

* * *

Bill Banerd was the first man hired to operate the Stewart airstrip. When Granduc started exploration at Leduc in 1954, one of the first orders of business was to build an airstrip, which they accomplished on cleared land adjacent to the Bear River. Aircraft with skis could land on the glacier near the mine pretty much year round but at Stewart wheels were needed for part of the year. Bill also operated the radio communication with the mine, and the company depended on him to hire men as ordered, and that's where I came in. Bill couldn't leave the radio unattended so if the mine told him they needed a carpenter for two weeks he would phone and ask me to rustle up a candidate. Some of the old vets were good enough men if you could get them out of town and dried out. Bill had a soft spot for these types, as did I. But some didn't work out too well at all, and I am reminded of the time Bill asked for a cook on a short-term job. The only person I could find was old Teddy, who insisted he could do it. As Bill told me later, the men asked Teddy to make them some pies. He mixed up enough crust for four pies, added filling and baked them. The four pies were set out, and one piece of each removed that first evening. Then they sat there in the same state for many days on end, until somebody threw them to the ravens. Teddy was never again asked to bake a pie. The crew said the ravens did survive.

* * *

About that time three or four young Newfoundlanders came to work at Granduc. It was seasonal work then but the young fellows thought they could make enough during a summer so that they could return to "the rock" for the winter. They were fine hardworking young men and I got to know them well. Every spring, Bill would have a job for the boys and if they had to wait for awhile for the mining job to start, I would give them room and board on credit. In the fall, they always paid me. One spring, the boys arrived only to learn that exploration would not proceed that year. They weren't the type to lay around so I paid their $300 fare to Vancouver. The following spring the boys didn't show up, and so I chalked up the loss of the $300 to experience. But four years later, I got a letter postmarked Baffin Island, Northwest Territories. I didn't know anyone there so it was with great curiosity that I opened it. Inside was a money order for $400 and a note from Junior Ryan, one of the Newfies, thanking me and telling me to buy a round for the boys!

BILL BANERD WAS THE SUPERVISOR OF THE STEWART AIRSTRIP AND A PART-TIME TRAPPER.

* * *

Guy Lawrence was the government telegraph operator at Stewart when I was growing up. From 1910 onwards, there was a single wire strung from Terrace to Alice Arm, under the inlet to Anyox, then up through a pass to Portland Canal at Maple Bay and then 40 miles alongside the canal to Stewart. The wire was strung from

➤ THE YUKON TELEGRAPH LINE WAS RUN FROM HAZELTON, BC TO DAWSON CITY AND ONE SECTION RAN BEHIND STEWART ALONG THE BELL IRVING RIVER AND THROUGH THE NINGUNSAW PASS. THESE ARE TWO LINEMEN AND THEIR PACK DOG. LEN MASON AND JOHN BARKER AT CABIN NUMBER EIGHT. SEE MAP AT START.

TELEGRAPH LINE. TREES WERE USED AS POLES WHEREVER POSSIBLE.

tree to tree and required two patrolmen with a gas boat to keep it intact due to storms and snowfall. But it worked most of the time and was abandoned only in 1938 when radio signals became reliable. When I was in the army cadets in 1940, Guy taught us Morse Code. He also wrote a book of his experiences on the Yukon Telegraph Line that ran from Hazelton, B.C. to Dawson City, Yukon Territory. The book, entitled Forty Years on the Yukon Telegraph, is a fine read but is now out of print.

* * *

John Hutchings acquired his nickname, Honest John, when he was Chief of Police at Anyox during the days of World War I. He was known as someone who could not be corrupted or paid to look the other way. In the twenties, he came to Stewart and with his wife operated the Bayview Hotel until the forties when he retired to Victoria. The ground floor of the hotel included the Burns Meat Market that was managed by Mike Murphy.

One of the Hutchings sons was Ozzie, who was a liquor vendor until 1935. Ozzie did some writing about the area, including the now out-of-print book, Stewart, the B.C. Mining Town that wouldn't Die.

The other Hutchings son, Bob, became my pal. He is still going strong at 80 and living in Stewart. I can't tell any wild stories about Bob in his youth because he says he knows even more stories about me!

* * *

When I was growing up in Stewart, it seemed that every second man was a veteran of World War I who was very proud of Canada and what the Canadian army had accomplished. One such man was Jimmy Burnett who was born in Scotland and with his family had moved to Winnipeg in 1912. When the war started, Jimmy and his father joined the Cameron Highlanders and went overseas in the same platoon as his father, who was the platoon sergeant. Sgt. Burnett, who had earlier won the MM (a medal for bravery), was leading his troops over the top in 1917 when he was killed. Jimmy was by his side. It always amazed me that when Jimmy would reminisce about the war, he would say, "If I had to do it over again, I'd never do anything different."

The Scotch Brigade was the name given to a group of Scottish vets who were mainly too old to work and just hung around town. They all had a thirst for the barleycorn and consumed such whenever possible. Bill Asselstine, who had been the Premier mill superintendent and was later elected MLA for Atlin Riding during the 1930s, had just arrived in town. The Brigade was hanging around the lobby.

"Hello, Bill," one said, "take us into the bar and buy us a drink."

Bill patted his own rear end. "Kiss this, boys, I kissed yours plenty and bought you drinks before the election. It's your turn now!"

* * *

Edgar Provinse ran the general store in Hyder, Alaska until 1944. He sold everything there, including gasoline, but liquor was his best seller. Alaska liquor laws were quite lax, and in those early days until now, there has never been a permanent police presence in Hyder. When Canada entered the war in 1939, liquor rationing went into effect but Edgar had unlimited supplies so his store became very busy. Canadian fish boats were regular visitors at the Hyder dock and many cases were shipped southbound to help the war effort. By 1944, Edgar had it made and headed south himself to retirement in Montana.

* * *

"Greasy Pete" Collins was a French Canadian who came north to work for me. He was a fair cook but someone started calling him 'Greasy' and it stuck. He would say, "I don't know why they call me Greasy because I only use the butter."

In the Granduc heyday, which started about 1969, Pete and a partner Henry started a furniture and appliance store. Business was very good and after one busy Christmas season, the store was quiet so Pete told Henry he was going out for a combined holiday and buying trip. He emptied the company account and with his wife and two young children booked a flight out on Pacific Western Airlines. Henry drove them to the airport and thought nothing of it when they loaded 23 suitcases on the plane. But Pete never returned and was never heard from again. Poor Henry was busted but to his credit, he continued operating and honored all the debts that Pete had piled up.

* * *

Kate Ryan was known as the "real" Klondyke Kate and was Sergeant Kate to all the sourdoughs. Kate moved to the Yukon in 1898 and spent some years as the police matron for the Royal Northwest Mounted Police at the border of Alaska and Yukon. Her duties included looking after female prisoners.

In the twenties, she moved to Stewart where she was government agent and the mining recorder for a time as well as matron of the hospital. Kate was also a businesswoman, the owner of a commercial building that is still standing on Fifth Avenue — Stewart's main street. Kate never married but raised her two nephews, Charlie and Johnnie, who became well-known horse packers in the area.

Kate passed away in Vancouver in 1934.

* * *

In the forties and fifties mining and construction companies operating in isolated areas expected their workers to spend ninety days at camp before being allowed to go home for time off. The time off was called the "turnaround". Joey was a machine operator who was a good man but he found ninety days away from his sweetheart very difficult so after a month or so he would quit the job so he could go to town. These types were known as "short stake artists". But Joey always had to have a reason to leave the job. My friend met Joey leaving camp with his packsack on his back.

"Did you quit again, Joey?"

"Yeah, I quit."

"What is the reason this time? It can't be the grub as the food here is first class."

"It's the food."

"How can that be?"

"The bloody hotcakes are too round!"

"Good bye Joey"

* * *

Henry was a Dane and he had a friend Pete who was also a Dane. Now Henry used to say that the most important trait that Danes had was that they were never known to tell a lie but after the following incident I never heard him repeat that saying.

Now Henry met Pete and was complaining that he wanted to go halibut fishing but could not procure any herrings for bait. Pete replied, "I used to fish lots for halibut at Alice Arm and we always used potatoes for bait."

Well, as Henry was hungry for fresh fish he got hold of ten pounds of potatoes and baited

COOKHOUSE AND CREW
SUNSHINE CLAIM ON GLACIER CREEK, 1909.
JOHN STEWART HOLDING DOG.

the fifty hooks on the halibut set line, then laid out the line off Bulldog Creek which was a known location for halibut. He let the skate of gear stay out overnight and went out the next morning to pull up the fish. No fish – the potatoes did not even show that they had been nibbled on.

So Henry went after Pete, "Pete, you told me a lie, the potatoes did not work."

"That's funny," Henry replied, "we always got halibut with potato bait. You did peel the potatoes first, didn't you?"

Only then did Henry see the twinkle in Pete's eyes and realize he had been had.

* * *

Owney McFadden was a packer and a prospector who owned Dempsey, a large half wolf, half malemute dog. In the summer months Owney used Dempsey as a pack dog and said Dempsey easily carried fifty pounds on his back all day.

One time in 1936, Owney was prospecting in the Meziaden Lake area and had set up his tent, cooked supper and crawled into the sleeping bag at dusk. Some time later he heard wolves howling on the sidehill but as that had happened many times he was not worried about Dempsey. But in the morning Dempsey was gone and although Owney stayed at the camp for a few days there was no sign of the dog so Owney went off to town.

Several months later it became known that a large beat up dog had been found at Alice Arm, sixty miles south by glacier from McFadden's camp. It turned out to be Dempsey and arrangements were made to have him returned to Stewart on the weekly steamship. Prior to getting lost Dempsey had fathered a brood of pups, one of which was given to the author. We called our pup Smokey and although he was gentle with us kids he did have a wild streak and loved to fight other dogs. By the time he was fully grown he was in trouble all the time and eventually got into John Campbell's pig pen and killed a few. Dad decided he had to go. He was given to a Native trapper who reported next spring that the wolves had lured him into the brush and killed him. There is no doubt that the wolves tried to do the same with Dempsey but that dog was too tough even for a pack of wolves.

"NO ORIENTALS WANTED." THIS WAS THE SAD MESSAGE ERECTED JUST UNDER THE STEWART BANNER (TO THE LEFT IN THE PHOTO) BY THE STEWART CITIZENS ASSOCIATION. BILL TOOTH, A LONG-TIME MAYOR AND LIBERAL ASSOCIATION KINGPIN, THOUGHT IT WOULD BE A VOTE-GETTER. WHEN A GROUP OF YOUNG SOLDIERS RETURNED FROM THE ARMY IN 1945, WE THREW THE SIGN INTO THE CANAL. BILL HAD A REPLACEMENT MADE, BUT WHEN WE ALSO TOSSED THE SECOND ONE OVERBOARD, HE GAVE UP. THIS PHOTOGRAPH WAS TAKEN IN 1943 AT THE STEWART DOCK DURING A CANADIAN SCOTS RECRUITING CEREMONY. AUTHOR'S MOTHER AND SISTER ISABEL IN FOREGROUND.

Chapter Thirty-Five

VILLAGE BUSINESS

Stewart was incorporated as a village in 1933. In 1967, I was elected mayor and held that post until 1982. Those were the Granduc boom years, and it was very satisfying to see our budget grow from $25,000 in my first year to $2 million in the final one.

It was the mayor's job during the first meeting of Council to appoint committee chairmen for various functions such as public works, fire department, recreation, etc. There were four aldermen plus myself, and we thought we were all old friends. The five of us had chosen various responsibilities but there was one post left — the hospital board. I called for volunteers. No takers. So I said, "Listen boys, we'll draw match sticks and whoever gets the shortest one has to take the job. But we mustn't tell anyone about this." They all agreed so we took five long matches, cut one short, and picked. I chose the short one and won the seat on the hospital board.

THE FIRE HALL BUILT IN 1910, WITH THE VILLAGE OFFICES UPSTAIRS. THE BELL TO CALL VOLUNTEER FIRE FIGHTERS WAS LOCATED IN THE TOWER, WHICH WAS ALSO USED TO DRY WET HOSES AFTER A FIRE.

The next morning I went to my coffee shop for breakfast and hadn't even sat myself down when someone yelled across the room: "Hey, Mac, I heard you got the short end of the stick last night!" I never again attempted secrecy during my terms as mayor.

When I started, the Village had two part-time employees — the garbage man and Alf Burton, who was a half-time village clerk plus part-time government agent, citizenship judge and sometimes custom's officer. Alf was a fine old Englishman who didn't like to be called that! He would say, "Don't call me an Englishman. I left there to get away from the English. I am a Canadian!"

He told me a story about one of his experiences as a citizenship judge. The immigrant applying for citizenship was required to know about Canadian government. Alf asked the applicant:

Who is the prime minister of Canada?

That would be Mr. Trudeau.

Who is the premier of BC?

That would be Mr. Bennett.

Who is the mayor of Stewart?

That would be King Eddy.

Since that was the well-worn nickname of my hotel, the King Edward, Alf gave him his citizenship!

In 1968, the Village became a District Municipality and expanded its boundaries to include the Granduc facility for tax purposes. With those additional funds, we were able to build an ice arena, a swimming pool, and a new combined municipal office/fire hall as well as to pave the streets, construct a sewage treatment facility, etc. The Granduc company had already built a water and sewer system and was renting that to the municipality. Eventually the company simply donated it to us.

One of the more interesting projects during my mayoralty was bringing live television to Stewart. In 1968, with the help of the Granduc company, we built a system of three mountain-top repeaters to carry the signal from Terrace. It had to travel about 120 miles and over some very tall mountain peaks. Most of the work was undertaken by local volunteers with the assistance of helicopters. The repeaters were powered by batteries that had to be replaced annually. It was a great day for the townspeople when the black-and-white signal was turned on, and it hardly mattered that the picture was so incredibly "snowy" we could barely recognize the puck during the NHL hockey games! Eventually the signal was upgraded and color became available. In the mid-70s, the municipality installed satellite receivers and broadcast six channels to the village. This system was later replaced by a private cable company, which continues to provide service although many residents prefer their own satellite receiver.

During my years on council, I was blessed with several fine crews of aldermen. We all seemed to have the same goal, which was to make the community a better place and especially to provide some decent recreational opportunities for our young people.

One of my long time fellow counselors recently reminisced about our days on council and how she had overheard the clerk say to me

INSTALLING TV REPEATER ON PEAK OF MOUNT EAST GEORGIE, 5,500 ELEVATION 1968

regarding some new counsellor: "For the first six months, they never say a word but after that they never stop talking."

Our family was discussing my retirement at the dinner table one night. My wife Vivian said, "It will be the end of an era, Ian." To which my son Robert cheekily added: "More like the end of an error!"

ST. FELIX CATHOLIC CHURCH. BUILT 1925.

Chapter Thirty-Six

HOSPITALS

STEWART'S FIRST HOSPITAL, BUILT IN 1910

The Stewart General Hospital, where I was born in 1927, was built in 1910. In the days before Medicare, it was often a financial challenge to find operating funds, and so it was closed temporarily on several occasions. Most years there was a doctor stationed in Stewart but not always. In such cases, Stewart residents would visit the doctor at the Premier mine, which was obligated to keep a physician on staff.

In 1953, Dr. Bill Hick was engaged as the local physician. Bill was and still is a mover and shaker of the first order. He got the locals organized, raised money and obtained grants so that a new hospital could be built. Bill eventually moved to Prince Rupert where he spent the rest of his medical career. Naturally, he continued his civic-minded ways and became mainly responsible for convincing the federal government and investors to build the new grain port in Prince Rupert. I was at a public meeting in Rupert when Pete Lester, the mayor, introduced Bill as "our grain surgeon."

THE PRESENT STEWART GENERAL HOSPITAL WITH MOUNT DOLLY IN THE BACKGROUND.

Dr. Hick's Stewart hospital was the victim of an accident in 1996. A machine operator was clearing snow from around the propane storage tank when he accidentally broke the line going into the building. Propane quickly filled the crawl space under the building, creating an immediate danger. But there was enough time to evacuate the staff before the building exploded. The ensuing blast lifted the hospital up and when it settled down again, it did not exactly match the foundation. Amazingly, there was no fire. However, the building could not be made safe again so a temporary hospital was established in a nearby building until the impressive modern new structure was built.

It is located on the same site as the two hospitals before it on Brightwell Street.

Chapter Thirty-Seven

SCHOOLS

STEWART'S FIRST SCHOOL IN 1910. TEACHER J.A. ROWE.

STEWART SUPERIOR SCHOOL IN 1935.

The Stewart Superior School, built in 1910, provided an education for grades one through twelve.

When I attended in the '30s and '40s, it had three rooms with a teacher for each room. In 1939, there were 100 pupils in total but that number diminished as people left the area for the war effort. It was quite primitive with outhouse-type toilets and a hand-pump. But we learned to read, write and count, and most of us did quite well in adult life.

In 1940, Mr. Cameron, the principal, left and his replacement was not qualified to teach high school — just in time for me to have to take correspondence! Most of the boys, like myself, found correspondence quite difficult but the girls seemed to get by somewhat better.

Hyder had a school during the 1920s but by

A GANG OF STUDENTS IN 1937.

1930, their children were coming across the line to attend the Stewart school. The State of Alaska paid their tuition. Recently Hyder opened its own school again.

Premier also had a school that boasted about 40 students in grades up to eight. The school board in 1935 adopted a policy to hire only male teachers for Premier. They said, "It was found that young women took better care of the children's school lives, while cupid took better care of the young women's lives." With 200 single men in camp, the single female teachers didn't stay single for long!

Stewart Superior School was replaced with a new modern structure in 1967 and became strictly an elementary school in 1975, when a separate high school was opened.

Chapter Thirty-Eight

INTERNATIONAL DAY FEATURES BRIDGE BURN

OLD-TIMERS AT INTERNATIONAL DAYS PARADE GROUND IN 1935. FROM LEFT, BILL TOOTH, BILL SMITH, AND OWNEY MCFADDEN, WITH HOOT MON ORR AT REAR LEFT.

The bridge over the Bear River is the only way into — and out of — town. In 1980, during the community's biggest annual celebration, the old wooden bridge burned to the ground. Not only did it extend the International Days party a little longer but it also ensured national media coverage.

Since the earliest days, Stewart and Hyder have celebrated their respective national holidays, July 1 and 4, together, and called the affair International Days. There are always parades, sports and various activities, and after road access arrived in 1972, many people from nearby communities like Terrace and Smithers would visit for the long weekend. On July 1, 1980, the town was full of visitors when somehow the Bear River Bridge caught fire and burned completely. Since there was no other way over the river and out of town, everyone had to stay.

Here are some excerpts from Maclean's Magazine's July 14, 1980 edition: "Trapped with the locals were about 1,000 tourists celebrating Stewart's 75th birthday. Yet nobody seemed terribly concerned. Most of the visitors refused free helicopter and bus rides to the outside. As Mayor

INTERNATIONAL DAYS PARADE ON FIFTH AVENUE IN 1958.

INTERNATIONAL DAYS PARADE, 1974

Ian McLeod explained, 'A lot of them had a good excuse to take another week off work without getting fired.'...

"The weekend before the July 1 holiday, the townspeople had started partying early with baseball games and a dance on Saturday. Then, on Sunday, celebrants took a casual drive across the border — unmanned on the American side — — carrying rifles for a turkey shoot in Hyder, Alaska... On Monday, most of Hyder was in Stewart, watching more ball games when, in mid-afternoon, the 90-metre wooden bridge caught fire. Within minutes, about 15 members of the Stewart volunteer fire department had turned out, but the creosote-soaked span collapsed before the three-hour blaze could be contained.

"Northern resourcefulness instantly surfaced. Mayor McLeod, trapped outside town

STEWART'S FIRST SKATING RINK AT BARNEY'S GULCH. BILL SMITH WAS THE BUILDER. HYDRO ELECTRIC PLANT TO RIGHT.

with the occupants of 20 cars, radioed for the three resident helicopters to carry them back into Stewart....Meanwhile, the provincial highways department reacted quickly by borrowing a Bailey bridge from a nearby river and ordering another to be brought by barge 180 km north from Prince Rupert...Local longshoremen worked without pay on their holiday to unload the bridge and haul it to the river.

"Almost everyone had planned to stay in town for the birthday party July 1, with its parade, beer garden and 300 pounds of barbecued salmon. But many of the visitors had expected to be leaving the next day. With the bridge still out, they had to improvise: (people) camping in Rainey Creek Park appointed a mayor, a minister of the environment (who complained to the municipal clerk about a plugged toilet) and a minister of fisheries — a woman who had caught a 50-pound salmon for a communal feast. The town supplied campers with firewood and a water truck and opened a swimming pool's showers to tourists. While supermarkets ran out of milk and one restaurant had to fly in fresh bread, the crucial shortage was

INTERNATIONAL DAYS PARADE. 1925.

booze: the liquor store was shuttered on its normal closing day, Monday, and again the next day for the holiday. Serious drinkers patronized town pubs and when those closed they attended Hyder's three all-night taverns where the indigenous drink is Snakebite — 190 proof grain alcohol downed in one gulp.

"Little wonder that late Thursday, when the temporary bridge over the Bear River again linked the town the rest of the world, few of the folks who had been stranded in Stewart seemed in much of a hurry to leave."

Investigators never discovered a cause for the fire, and arson was not suspected. However, the day after the fire, I overheard someone muttering: "That Charlie Fenner (owner of Bob's Mercantile store) probably set the bridge on fire so he could sell all his groceries to us!"

OLD SWIMMING HOLE. CLEMENTS LAKE.

Chapter Thirty-Nine

WARD'S PASS CEMETERY, STEWART

WARD'S PASS CEMETERY, STEWART

The first cemetery in Stewart was located on the east side of Bear River about a thousand yards above the Bear River Bridge. About six wooden slabs memorialized people whose names were not familiar to me. When I last saw it, I was about ten years old and the river was fast eroding the land beside the cemetery. When I tried to find the place in the mid-1940s, everything had been washed away.

In the late 1920s, the town established another cemetery on the valley floor beside Barney's Gulch Creek, located two miles outside of town. It was a well-organized place, with special sections for members of the local Legion, Moose Lodge, and Masonic Lodge.

Several times in my youth, I overheard Mother tell Dad in her broad Scottish accent: "John, if I die don't let them put me in the waters." When I asked what she meant, her reply was always that I was too young to understand.

One summer day, when I was about twelve, my pals and I were out riding our bicycles when we noticed a funeral procession leaving the Moose Hall. We followed them on the mournfully slow drive out to Barney's Gulch and then stood back under the trees to watch the graveside service. The newly dug grave was covered with a tarpaulin, and the coffin was set on two planks above the tarp. The service was completed, the mourners went home, and we young lads moved closer to watch as the two gravediggers finished the job. One fellow, swinging his pick, punched a couple of holes into each end of the coffin. They looped ropes around the coffin and then lifted it up, asking us boys to slip the planks out from underneath. As we did, I glanced down into the hole and saw that it was about one-third full of dirty grey water. It struck me at that instant what Mother had meant: she didn't want to be buried here! (The holes in the coffin, of course, were to help it sink into the water.) When Dad died in 1952, there was no argument that the body be transported to and then interred in the Burnaby Masonic Cemetery. Thirty years later, Mother was laid beside him. No, Mom, you're not "in the water!"

In the late fall of 1958, the first snowfall of the winter, about a foot deep, was followed by days upon days of warm rain. Flooding was everywhere, and the steep banks of Barney's Gulch, upstream of the cemetery, sloughed into the little river and dammed it up. Nobody knows how much water built up behind the natural dam but when it finally let go, millions of gallons of rushing water came barreling down the gulch and transported millions of tons of gravel to the valley floor. The cemetery was completely buried under 30 or more feet of gravel, and that was the end of Barney's Gulch Cemetery.

In 1960, Stewart citizens cleared an acre of land adjacent to Ward's Pass, about five miles from Stewart, for use as a cemetery. The Pass had been named after Ward Brightwell (whose name is also on Brightwell Street), a pioneer from the turn of the century.

When Ward lived in the area, a big bluff existed just above the cemetery site, and the Bear River flowed against the bluff. Ward cut a trail over the bluff that was later named for him. The trail became a wagon road, then an automobile road, and in 1947 the bluff was blasted out of existence to make way for the present roadway.

RIVERVIEW CEMETERY, HYDER, ALASKA

The cemetery at Hyder, Alaska is located just past the present waterworks plant. There are not many graves as Hyder was never as populous as Stewart. It was originally known as the Riverview Cemetery but, as a second growth of trees have since matured, the river is no longer in view.

Just above this cemetery was the Moose Park. The Moose Lodge was an international fraternal organization and its local members dammed off a small creek to make a swimming pool for the youth of the '30s. There were picnic tables and a diving board, but it all went back to nature when the men went off to war.

Chapter Forty

GRIZZLIES AND FISH

FISH CREEK HOUSE IN 1935

Fish Creek has always been an important landmark in the Portland Canal area. This lovely creek, located four miles north of Hyder on the road to Granduc, is the spawning ground for thousands of Chum Salmon every August through October. That also makes it an important feeding place for local grizzly bears. The U.S. Forest Service has recently

THE GRIZZLY BEAR VIEWING PLATFORM, 2003

erected excellent bear viewing platforms for the thousands of tourists who come to watch the antics of the grizzlies as they fish.

In 1922, Helen Bell and Elma Smith built a roadhouse close to the creek and called it the Chicken Inn (pictured here). They planned to serve chicken dinners with country gravy as well as sundry other services and goods. Business was good until the salmon arrived on their annual upstream journey, spawned and died — almost right outside the door of the Chicken Inn. The odor of decaying fish was quite strong and not terribly conducive to romantic chicken dinners! The venture soon failed.

When I was a boy in the 1930s, Sid Anderson and his wife lived in the Chicken Inn, which was later referred to as Fish Creek House. Sid was the United States Commissioner and he performed marriages for dozens of Canadian couples. Canada had a legislated waiting period between the license and the ceremony, but not so in the U.S.A. Many miners, who had very few days of leisure during the year, preferred the quicker method!

The flood of 1961 destroyed the house, which had stood vacant for many years. You may still be able to see the fireplace standing resolutely in the bush west of the present grizzly bear viewing area parking lot.

➤

THE U.S. FOREST SERVICE ADMINISTERS FISH CREEK.

➤

YOUNG GRIZZLY FISHING AT FISH CREEK. PHOTO TAKEN FROM VIEWING PLATFORM.

SS CATALA OF UNION STEAMSHIPS AT STEWART DOCK, 1935

Chapter Forty-One

THE STEAM SHIPS

SS AMUR AT CRAWFORD DOCK, 1930. OLD RAILWAY TRESTLE AND ROUNDHOUSE ARE IN BACKGROUND.

In the twenties and thirties, the two communities of the Portland district were served weekly from Vancouver, B.C., by Union Steamships and Canadian National Steamships.

Since water was the only access into the district, these ships carried with them virtually all supplies including supplies, men and equipment. When I built the present King Edward Hotel in 1971, the lumber, plywood, and all other construction materials were shipped from Vancouver via ship.

On their six-day return trips, Union generally used the ship "Catala" while CN employed the "Prince George." Both featured fine passenger accommodation, excellent dining rooms and room for a great deal of freight. When the Prince George entered the harbour, she not only blew her whistle but also ran her siren. In the winter, the wolves on the hillside would then start howling in response!

In the 1960s, Northland Navigation assumed the operation of Union Steamships and continued good service to the coastal communities. Folks lucky enough to sail on her were treated to an experience and views that easily equal the present-day Alaska cruise boat excursions. The "Northland Prince" was a fine ship with 50 staterooms and a large dining room. On the last trip before Christmas, they always hosted a party for the townspeople in a decorated dining room with gourmet food and fine wine. The event was classed by the locals as the affair of the year, and everyone dressed up in their finest to associate with the ship's officers who were naturally in full uniform.

The arrival of the Northland Prince every Saturday morning was the highlight of the week for most residents, and shortly after its' docking there would be a mad dash to Bob's Mercantile grocery store to buy the newly-arrived fresh fruit and produce. If one was a little late that day, you could be out of luck for the entire week!

The Liberal government of Canada discontinued the subsidy for the Northland route in 1976, sounding the death knell for many coastal communities such as Alice Arm. Stewart, fortunately, had just recently established road access to the rest of the province, but the steamship service was still sorely missed.

SS PRINCE ROBERT AT THE SECOND STEWART DOCK IN 1925.

SS PRINCE ROBERT AT THE FIRST STEWART DOCK THAT WAS LOCATED AT THE SITE OF THE PRESENT YACHT CLUB. NOW THE WATER IS TOO SHALLOW FOR SUCH A LARGE SHIP. IN THE FOREGROUND IS THE STEWART ROAD ON PILINGS WHICH WAS WELL USED UNTIL THE PRESENT ROAD TO HYDER WAS BLASTED OUT OF THE SIDE HILL. MY MOTHER SAID SHE OFTEN WALKED ME IN THE BABY BUGGY ON THE PLANK ROAD.

Chapter Forty-Two

BEAR PASS

Highway 37A follows a scenic route through the Bear Pass, some 20 miles north of Stewart. The highest point of the highway is today at only 600 feet above sea level, providing a gentle grade, but this wasn't always the case. Until natural glacial activity changed the landscape in the mid-20th century, the highway hung about 1000 feet above Bear Glacier and Strohn Lake, providing a spectacular but dizzying view!

The Bear Glacier (known until 1910 as Grosse Glacier: see map on page one) completely filled the valley until 1958 when its recession caused the phenomenon known as "jokulkaup." It is a chain reaction originally caused by a sudden buoyancy of the ice, an outburst of ice-impounded water, and the resulting outflow of water from lakes that were previously dammed by the glacier (in this case, Strohn Lake).

THE GLACIER AT BEAR PASS IN 1928. THE PRESENT ROAD IS NOW IN THE CENTER OF THE PICTURE. THE SMOKE IS FROM THE GEORGE ENTERPRISE MINING CAMP, SNOW LAKE IN DISTANCE.

BEAR GLACIER AT STROHN LAKE, 1999.

STROHN LAKE AND HIGHWAY 37A AT BEAR PASS. SMALL POND ON THE LEFT IS WHAT CONSTABLE LARRY REQUA SAID WAS CAUSED BY A METEORITE IMPACT IN 1938.

GEORGE COPPER MINING CAMP 1930 WITH THE BEAR VALLEY BELOW. BEAR PASS WOULD BE TO THE RIGHT.

The first jokulkaup scoured the valley floor, causing much damage to the road that was being constructed downstream and filling Snow Lake with gravel. The phenomenon reoccurred in 1959 and 1960. Because of the downstream damage, the provincial government drilled and blasted a channel in the ice to drain the water as much as possible. The water level in Strohn Lake was eventually reduced by 200 feet and it now drains into the Bear River.

In 1910, a horse trail had been built up high on the hill, on the west side of the valley above Bear Glacier. In the mid-1950s, it was upgraded to handle motor vehicles but it was abandoned in 1972 when the present low-level highway was built. (The old road remains visible, perched high above the highway.)

In the mid-1960s, when the west side of the glacier had melted sufficiently, the present highway was started as a logging road alongside the glacier. In 1972, it became part of a transportation system intended to tranship asbestos from Cassiar mine to the Stewart port and was financed by the "Roads to Resources" program of the federal Diefenbaker government. The road to Cassiar was opened in 1969 and the connection to south-

ern BC was completed in 1971 with the opening of the Nass Bridge.

In the early 2000s, the Bear Pass and Glacier were used as sets for the movie "Insomnia" starring Robin Williams and Al Pacino.

Although the Bear Glacier has receded across the valley, there is still a great deal of ice as it is connected to the Cambria snowfield, which reaches 60 miles south to the community of Alice Arm.

The glaciers in the Stewart area have been extensively studied by scientists, some of whom

▲ FRED YOUNG, JACK RENNIE, AND JOHNNIE RENNIE ON THE HORSE TRAIL AROUND BEAR PASS IN 1937. THE ROAD BUILT ON THE HILL MORE OR LESS FOLLOWED THIS HORSE TRAIL.

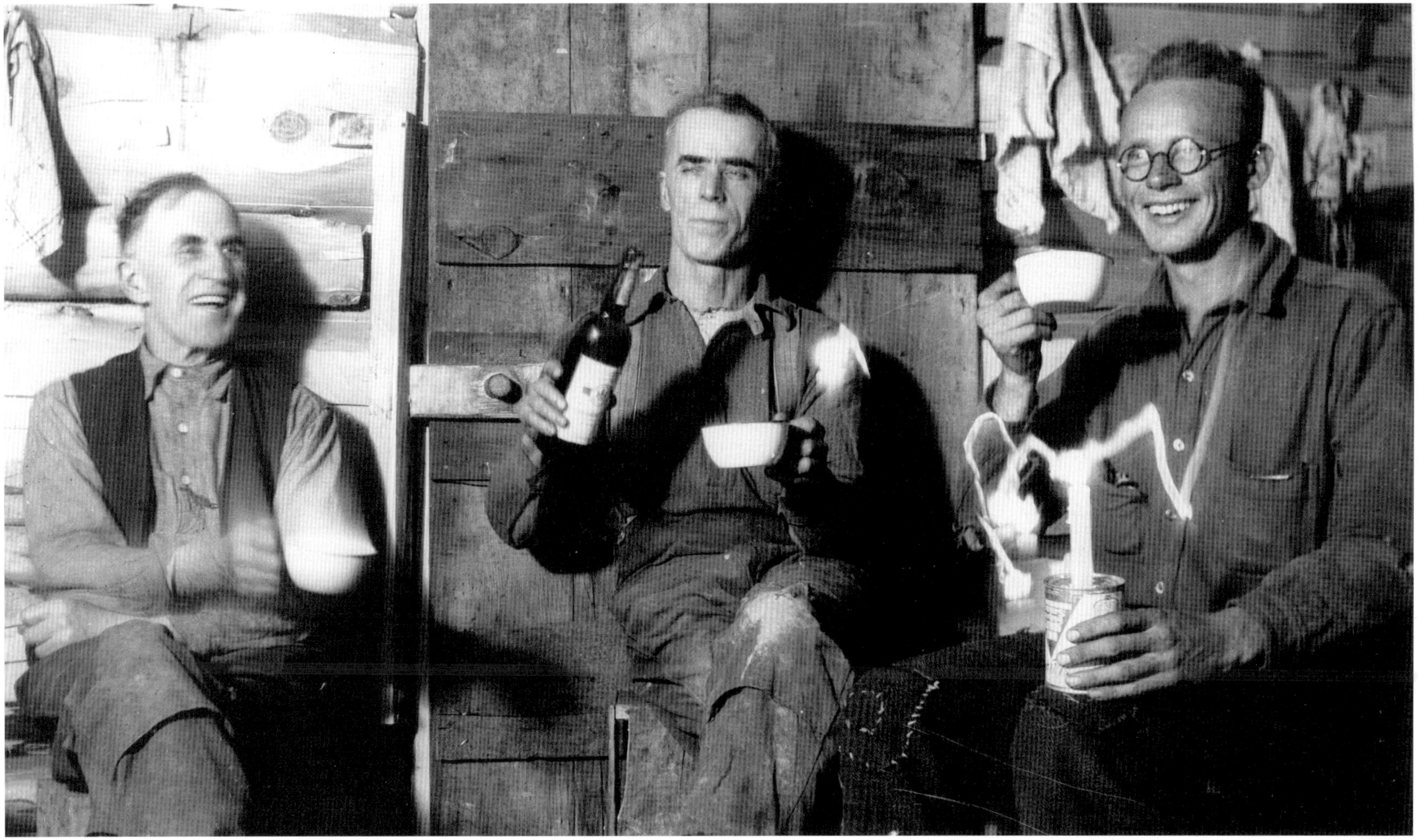

RIVA BROTHERS AND JOHN SUTHERLAND AT MONTREAL CLAIMS, TWO MILES EAST OF BEAR PASS, 1930.

have identified a species of resident ice worms (Latin name Mesenchytroeus solifugus). They abound specifically in the Cambria and Frankmackie snowfields, both in the Portland region, most often at elevations of 2,500 to 3,500 feet. They are black or brownish black, looking very much like an earthworm of less than one inch in length. They live mainly in surface sink-holes and melt-water channels, but have been observed moving from their channels onto clean ice surfaces in order to soak up bright summer sun!

THE NASS BRIDGE OPENED IN 1971, CONNECTING STEWART AND THE PORTLAND CANAL WITH THE REST OF THE PROVINCE ALONG THE NEW HIGHWAY 37.

➤ MEZIADEN FISHWAY BUILT JUST ABOVE WHERE THE MEZIADEN RIVER JOINS THE NASS RIVER.

EMPTY PACK TRAIN CROSSING BEAR GLACIER.